Springer Series on
SIGNALS AND COMMUNICAT

# Signals and Communication Technology

**Functional Structures in Networks**
AMLn – A Language for Model Driven Development of Telecom Systems
T. Muth
ISBN 3-540-22545-5

**Radio Wave Propagation for Telecommunication Applications**
H. Sizun
3-540-40758-8
ISBN

**Electronic Noise and Interfering Signals**
Principles and Applications
G. Vasilescu
ISBN 3-540-40741-3

**DVB**
The Family of International Standards for Digital Video Broadcasting, 2nd ed.
U. Reimers
ISBN 3-540-43545-X

**Digital Interactive TV and Metadata**
Future Broadcast Multimedia
A. Lugmayr, S. Niiranen, and S. Kalli
ISBN 0-387-20843-7

**Adaptive Antenna Arrays**
Trends and Applications
S. Chandran (Ed.)
ISBN 3-540-20199-8

**Digital Signal Processing with Field Programmable Gate Arrays**
U. Meyer-Baese
ISBN 3-540-21119-5

**Neuro-Fuzzy and Fuzzy-Neural Applications in Telecommunications**
P. Stavroulakis (Ed.)
ISBN 3-540-40759-6

**SDMA for Multipath Wireless Channels**
Limiting Characteristics and Stochastic Models
I.P. Kovalyov
ISBN 3-540-40225-X

**Digital Television**
A Practical Guide for Engineers
W. Fischer
ISBN 3-540-01155-2

**Multimedia Communication Technology**
Representation, Transmission and Identification of Multimedia Signals
J.R. Ohm
ISBN 3-540-01249-4

**Information Measures**
Information and its Description in Science and Engineering
C. Arndt
ISBN 3-540-40855-X

**Processing of SAR Data**
Fundamentals, Signal Processing, Interferometry
A. Hein
ISBN 3-540-05043-4

**Chaos-Based Digital Communication Systems**
Operating Principles, Analysis Methods, and Performance Evaluation
F.C.M. Lau and C.K. Tse
ISBN 3-540-00602-8

**Adaptive Signal Processing**
Applications to Real-World Problems
J. Benesty and Y. Huang (Eds.)
ISBN 3-540-00051-8

**Multimedia Information Retrieval and Management**
Technological Fundamentals and Applications
D. Feng, W.C. Siu, and H.J. Zhang (Eds.)
ISBN 3-540-00244-8

**Structured Cable Systems**
A.B. Semenov, S.K. Strizhakov, and I.R. Suncheley
ISBN 3-540-43000-8

**UMTS**
The Physical Layer of the Universal Mobile Telecommunications System
A. Springer and R. Weigel
ISBN 3-540-42162-9

**Advanced Theory of Signal Detection**
Weak Signal Detection in Generalized Observations
I. Song, J. Bae, and S.Y. Kim
ISBN 3-540-43064-4

**Wireless Internet Access over GSM and UMTS**
M. Taferner and E. Bonek
ISBN 3-540-42551-9

Djuro G. Zrilic

# Circuits and Systems Based on Delta Modulation

## Linear, Nonlinear, and Mixed Mode Processing

With 130 Figures

Prof. Dr. Djuro G. Zrilic
Department of Engineering
New Mexiko Highlands University
National Ave. 803
Las Vegas, NM 87701
U.S.A.

ISBN 3-540-23751-8 **Springer Berlin Heidelberg New York**

Library of Congress Control Number: 2005921900



**Springer is a part of Springer Science+Business Media**

springeronline.com


Printed in The Netherlands



Typesetting: Data conversion by author.
Final processing by PTP-Berlin Protago-TEX-Production GmbH, Germany
Cover-Design: design & production GmbH, Heidelberg
Printed on acid-free paper 62/3141/Yu - 5 4 3 2 1 0

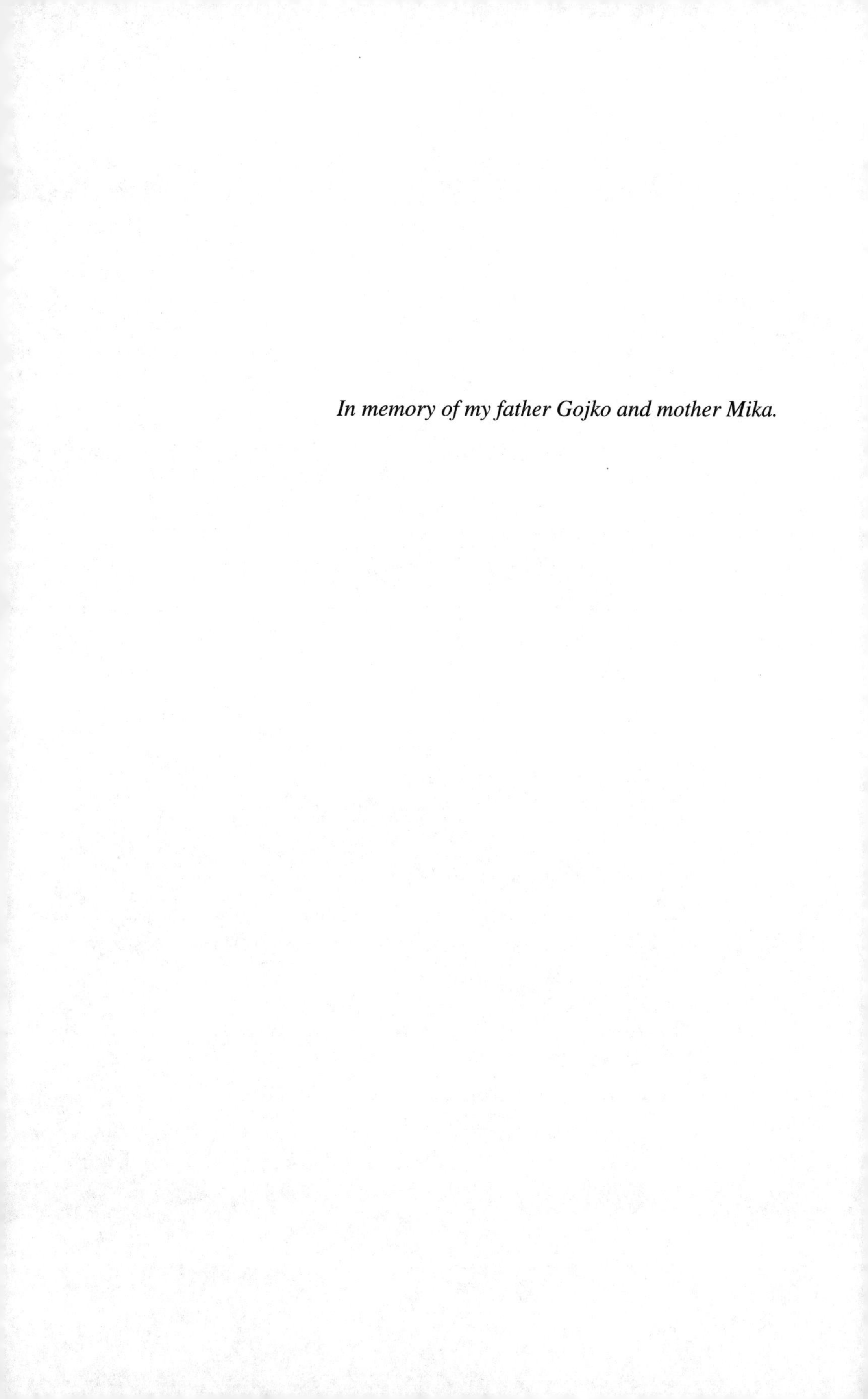

*In memory of my father Gojko and mother Mika.*

# PREFACE

This book has the distinction of being the pioneering work on the subject of arithmetic operations of multi-valued delta modulated pulse density streams. Its focus is on both theoretical and practical aspects of different applications of linear, nonlinear, and mixed mode processing. The idea stems from a digital differential analyzer (DDA). Intensive design studies of DDA were undertaken during the 1950s in the Soviet Union, the USA, and the UK. The underlying aim was to replace software with hardware to perform real-time operations. The operation of DDA is based on the method of delta-modulation, where the idea is to build a functional arithmetic unit whose input and output is a delta modulated pulse density stream consisting of both negative and positive pulses. The instantaneous analog values are obtained by continuously averaging (demodulating) the polarities of pulses while a particular problem is being solved. Depending on the problem, individual functional units are interconnected in a similar manner to that in analog computers.

This book presents a number of such functional units for linear, nonlinear, and mixed mode processing of delta modulated pulse streams. We hope that this text provides a basis for a complementary approach to real time signal processing where traditional methods are losing their practical and economical significance.

A number of people have contributed to this text. I am grateful to Prof. Georgie Lukatela, Prof. Grozdan Petrovic, Prof. Savo Leonardis, and Prof. Lewis Franks, who in my youth introduced me to the ideas of pulse code modulation, delta modulation, digital communication, and the digital revolution in general. I am also grateful to my colleagues, Dr. Rade Majkic, Dr. Vojin Senk, Mr. Milan Narandzic, Mr. Nebojsa Pjevalica, and Dr. Danilo Mandic for their valuable comments and suggestions. A landmark paper by Dr. Nik Kouvaras, which was published in 1978 had significant influence on my work. I would like to express my gratitude to Dr. Gilbert Sanchez, my former president, who is a great proponent of science and engineering. I would also like to thank Mr. Umesh Dole, who is a graduate student at the University of New Mexico, and who helped me generate the simulation results in the text. I would also like to thank Dr. Glen W. Davidson and Mr. Tim Ames of Santa Fe, who edited the text.

Finally, I wish to express my deepest gratitude for the steadfast support of the late Dr. Dan C. Ross and Dr. Tomislav Tomic.

Djuro G. Zrilic
Santa Fe, NM.

# INTRODUCTION

This book presents a collection of published and ongoing research work of the author over the past two decades in the area of delta modulation. Delta-sigma modulation (Δ-ΣM) is a very attractive, high resolution serial analog-to-digital converter (ADC). The output of a Δ-ΣM is a high rate, one bit, serial pulse density stream. There exists many different Δ-Σ modulation architectures. The bandwidth of an analog signal to be converted into a Δ-Σ modulator is much smaller compared to the bandwidth of the digital format of the Δ-Σ modulator output. Delta-sigma data converters exploit coarse quantization, which introduces quantization noise. This error is suppressed using negative feedback around the quantizer within some frequency band of interest. Δ-Σ ADCs are known as noise-shaping converters and they are characterized by high sampling rates, high resolution, and high signal-to-noise ratios (SNRs) over a wide dynamic range.

This book focuses on the processing aspects of a delta modulated pulse density stream. There exists two different approaches in this field. The traditional approach is to convert the Δ-Σ pulse density stream to a pulse code modulated (PCM) signal by means of decimation, and then process the multi-bit PCM word using ordinary digital signal processing (DSP) hardware. The second approach involves direct processing of a delta modulated pulse stream. To achieve this, a number of circuits have to be synthesized to perform linear, nonlinear and mixed mode arithmetic operations on a delta modulated pulse density stream. Due to over-sampling, the interface of a Δ-ΣM with the analog world is not expensive and is not complex compared to practical PCM interfaces. There are a number of applications where a Δ-Σ modulator is integrated with different types of sensors, thus eliminating the need for special types of additional interfacing circuits. Our goal is to add more functions on the same Δ-ΣM Integrated Circuit (IC). To implement this, additional novel circuits for direct processing of a Δ-Σ modulated pulse density stream have been developed.

In the following chapters, we have synthesized a number of circuits and systems for linear, nonlinear, and mixed mode processing of a Δ-Σ pulse density stream. The first three chapters of this book serve as a general introduction, and present a compilation of the existing literature. The first Chapter covers the basics of binary and multi-valued delta modulation. In

Chapter 2, some existing approaches of linear processing of a delta-modulated pulse density stream are presented. In Chapter 3, some basics of multi-valued logic are given. This introduction is needed for understanding Chapter 4, where the possibility of arithmetic operations on multi-valued delta modulation pulse density streams is examined. In fact, Chapter 4 presents a generalization of the existing binary arithmetic approaches performed on a binary delta modulated pulse density stream. In Chapter 5, we have shown the possibility of direct nonlinear processing on a delta modulated pulse density stream. In Chapter 6, the specific applications of a delta modulated pulse density stream are explored for mixed analog/digital processing. In Chapter 7, performance of two types of linear decoders are analyzed and compared to nonlinear decoders in the presence of channel errors. In Chapter 8, two different methods of direct conversion of a PCM binary word into a Δ-ΣM 1 bit pulse density stream are presented. In Chapter 9, two examples of stochastic processing of a Δ-ΣM pulse density stream are given. An example of using binary Δ-Σ arithmetic in measurement and instrumentation is elaborated in Chapter 10. Chapter 11 gives examples of compression of low-pass and band-pass Δ-ΣM pulse density streams. In addition, possibility of arithmetic operations on band-pass Δ-ΣM pulse density streams is demonstrated.

# CONTENTS

# CHAPTER 1 DELTA MODULATION SYSTEMS

## 1.1 LINEAR DELTA MODULATION SYSTEMS

Although invented in 1946 [1], delta modulation only received full recognition in the last decade or so. It did not gain importance until recent developments in the mixed analog-digital very large scale integration (VLSI) technology. There are several types of delta modulation systems, such as: linear delta modulation (LΔM), delta-sigma modulation (Δ–ΣM), adaptive delta modulation (AΔM), etc. This chapter is dedicated to the fundamentals of both LΔM and Δ–ΣM.

### 1.1.1 The Principle of LΔM

Linear delta modulation is a non-linear sampled data closed control loop system. It is well understood and described in the literature [2]. Let us consider a linear delta modulation system first. Fig. 1.1 shows a block diagram of a linear delta modulator and demodulator with belonging waveforms.

As can be seen from Fig. 1.1, an analog input signal is encoded by the delta modulator into a binary pulse stream. Delta modulation is based on a sampled binary quantizer, quantizing the change in the input signal from sample to sample. Output pulse stream $X_n$ is locally decoded back into an analog waveform by an integrator. The integrator acts as a predictor and attempts to predict the input $x(t)$. The prediction error, $\varepsilon(t) = x(t)\text{-}\hat{x}(t)$ is quantized and sampled in the binary quantizer (BQ). The negative feedback of the delta modulator ensures that the polarity of the pulses is adjusted by the sign of the error signal, which causes the locally reconstructed signal to "follow" the input signal. As a result, the binary quantizer produces the sign of the difference between the input and feedback signal. This difference is called "delta", hence the name "delta modulation" [2].

The modulation is called linear because the local decoder (integrator) is a linear network. In this chapter we will focus on linear local decoders. Delta modulator output $X_n$ is integrated (for errorless transmission) in the receiver just as it is in the feedback loop. The integrator produces a waveform, which differs from the original signal by the error signal in the modulator. By low-pass filtering (LPF), the majority of the quantization noise, which lies outside the message band, is removed.

### 1.1.2 Basic Parameters of LΔM

In section (1.1.1), we gave a qualitative description of the basic principles of LΔM. Now it is necessary to define the basic relationships between parameters relevant for proper operations of LΔM. To simplify analysis, let us assume that the input signal is a continuous and periodic function.

$$x(t) = A\cos(\omega t + \theta) \tag{1.1}$$

where $\omega = 2\pi f$, and $f$ is the frequency of the input signal. Suppose that the size of an amplitude quant is $\delta$= constant. The design question is how high the sampling frequency, $f_s = 1/T_s$ , needs to be so that the approximation signal (reconstruction signal) $\hat{x}(t)$ follows the input signal $x(t)$ correctly, i.e., that there is no slope overload,

$$|\varepsilon(t)| = |x(t) - \hat{x}(t)| < \delta\,. \tag{1.2}$$

In Fig. 1.2a, the step size is too small and because of that the feedback signal $\hat{x}(t)$ is not able to follow ("hunt") the rapid rise of the input signal. In this case, we say that the system suffers from the slope overload. This type of error is called slope overload error.

Correct "hunting" (tracking) is shown in Fig. 1.2d when the sampling interval is chosen properly. The error difference is now in the range $\pm\delta$, and its maximum value is $+\delta/2$ or $-\delta/2$ and it is known as quantization error. This error is irreversible and it is the price of digitalization. Throughout the book we will always assume that the system is correctly oversampled and that the slope overload never occurs. In addition, the problem of correct tracking can occur when the $\delta$ step size is too large, Fig. 1.2b. In this case the input signal is smaller than the step size and the reconstruction is not possible.

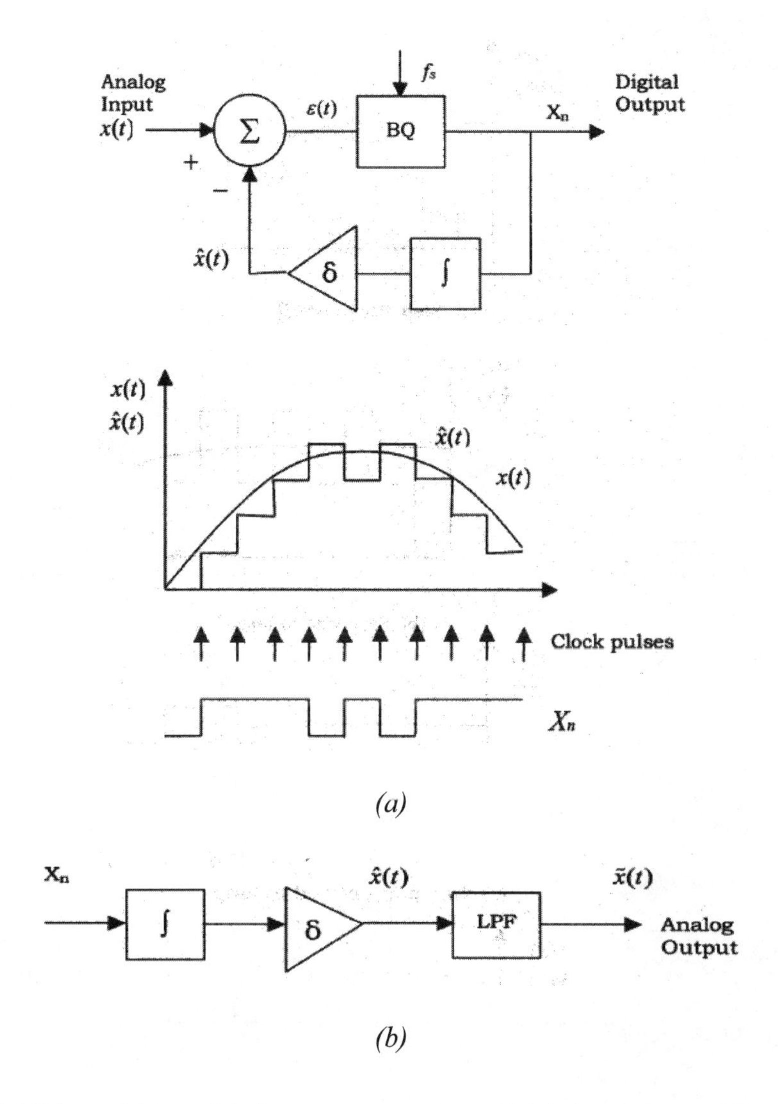


**Fig. 1.1.** Linear delta modulation systems, **(a)** LΔM with belonging waveform, **(b)** demodulator

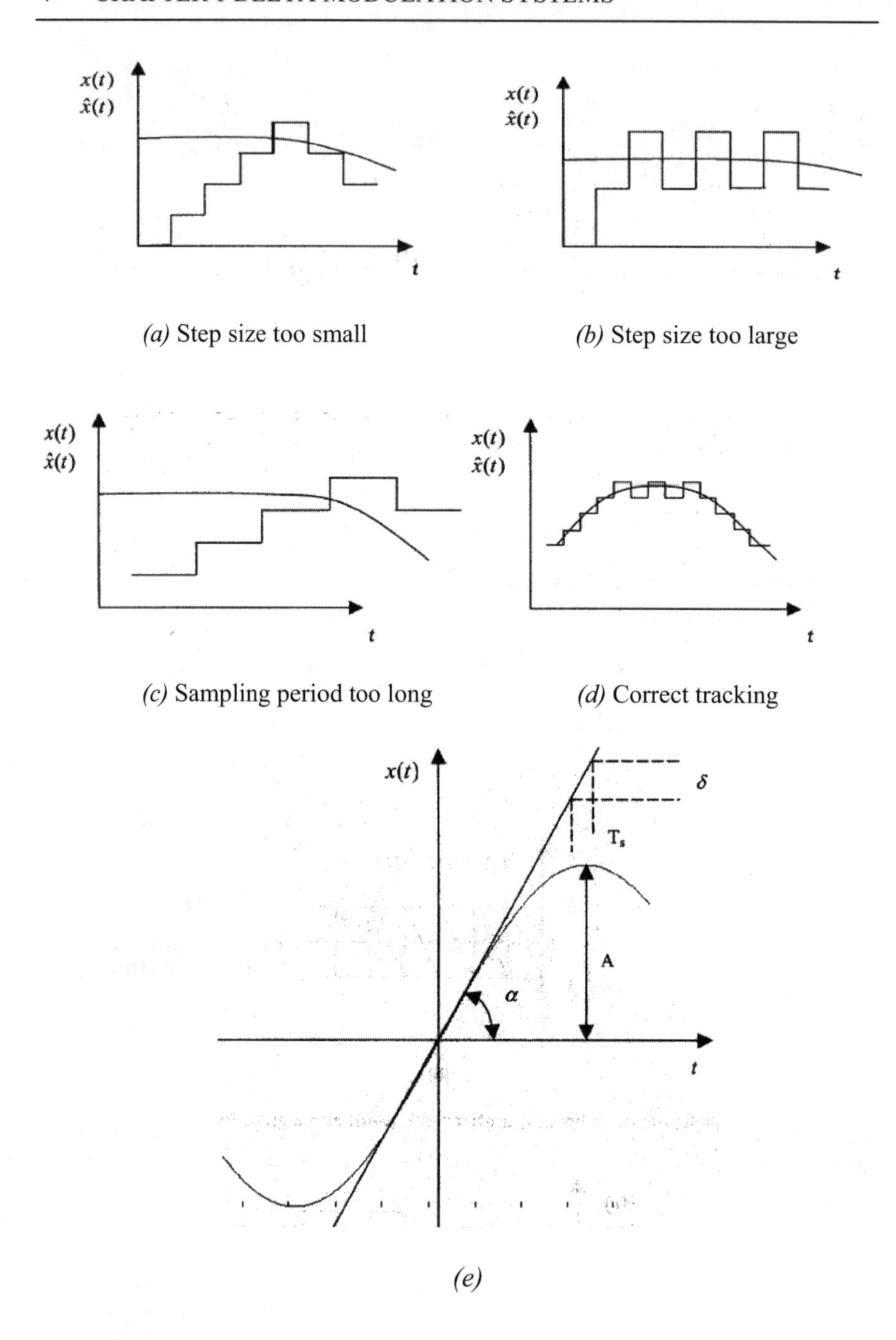


**Fig. 1.2.** Critical design parameters in a linear ΔM system

From the waveforms in Fig. 1.2, we can observe that correct conversion is possible if the maximum slope of the input signal does not overshoot the slope. The following condition has to be satisfied for correct tracking.

$$\max\left\{\frac{dx(t)}{dt}\right\} < tg\alpha = \frac{\delta}{T_s}. \tag{1.3}$$

For periodic signal (1.1) this condition is satisfied if

$$2\pi fA < \frac{\delta}{T_s} = \delta f_s. \tag{1.4}$$

To avoid slope overload, the maximum amplitude of the sinusoidal signal has to satisfy the inequality

$$A_{mx} < \frac{\delta}{2\pi}\left(\frac{f_s}{f}\right) \tag{1.5}$$

for fixed $\delta$ and $f_S$. This is a very important and interesting result, which tells us that the maximum allowed amplitude of the input signal to the LΔM decreases with the increase of the input signal frequency. This means that ordinary LΔM and some types of AΔM are suitable for A/D conversion of signals whose frequency amplitude spectrum decreases with increase of frequency. Human speech, for example, is such a signal. Using this property Motorola successfully produced and commercialized the so-called continuous variable slope ΔM system (CVSΔM, MC 3417) for telecom applications. The conversion has no slope-overload error as long as the eq. (1.5) is satisfied, and the resulting difference signal $\varepsilon(t)$ has a similar waveform as the quantization noise of Pulse Code Modulation (PCM). This error signal, commonly called the triangular wave quantization noise, is shown in fig. 1.3.

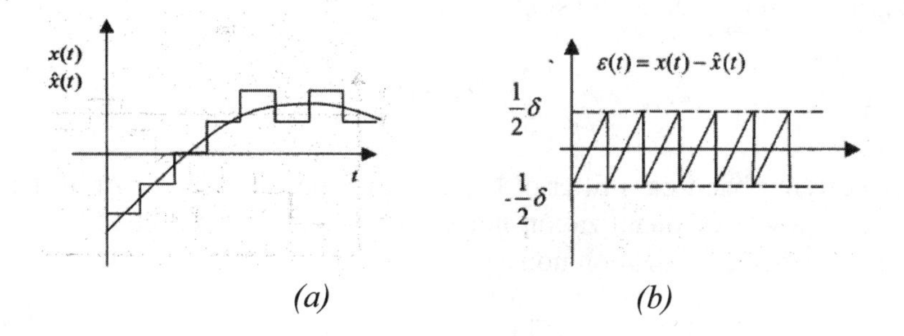


**Fig. 1.3.** Waveform of uniform quantizer, **(a)** quantized signal, **(b)** quantized error

We will use the linear model of the ΔM system in our consideration, and we will assume that the spectral power density of quantization noise is practically uniform, even thought this is not always the case [5]. Uniform power spectral density of the quantization noise means that the power of this noise will be directly proportional to the bandwidth of the receiver filter, i.e.

$$P_\delta \div B\,. \tag{1.6}$$

The number of spectral components of quantization noise in bandwidth B will increase as the sampling frequency is decreased:

$$P_\delta \div \frac{B}{f_s}\,. \tag{1.7}$$

Finally, increasing the amplitude of delta-step size, $\delta$, will increase the effective value of quantization noise of LΔM. Thus $P_\delta \div \delta^2$, or

$$P_\delta = K_\delta \left(\frac{B}{f_s}\right)(\delta)^2 \tag{1.8}$$

where $K_\delta$ is some constant of proportionality. In the case of the speech signal with quantization noise only (no slope overload), this constant is $K_\delta = 0.33$.

If it is assumed that the spectral power density of the quantization noise is strictly uniform in a given frequency bandwidth, and if we assume that instantaneous values of error signal $e(t)$ are uniformly distributed in the amplitude region $-\delta$ to $+\delta$, then it is possible to derive an analytical expression for the mean power of the quantization noise, and to define exactly the constant $K_\delta$ in eq. (1.8).

The unfiltered noise signal $P_\delta$, when the delta modulator is operating in the granular noise mode (no slope overload) is:

$$P_{\delta total} = \frac{1}{2\delta}\int_{-\delta}^{\delta} \varepsilon^2(t)\,d\varepsilon(t) = \frac{\delta^3}{3} \tag{1.9}$$

The receiving band-pass filter of bandwidth B passes only a part of the total mean power of quantization noise. For these idealized conditions, the mean power of quantization noise is:

$$P_\delta = \frac{1}{3}\left(\frac{B}{f_s}\right)(\delta)^2 \tag{1.10}$$

Comparing expressions (1.8) and (1.10) it can be concluded that the constant of proportionality $K_\delta$ is exactly 0.33.

The mean power signal-to-quantization noise ratio, when the input signal is a sinusoid of amplitude A is:

$$\rho^2 = \frac{P_s}{P_\delta} = \frac{A^2/2}{K_\delta \left(\frac{B}{f_s}\right)\delta^2} = \frac{1}{2K_\delta}\left(\frac{f_s}{B}\right)\left(\frac{A}{\delta}\right)^2 \tag{1.11}$$

We can see that the signal-to-quantization noise ratio increases linearly with the increase of the mean power of the input signal. At maximum allowed mean power the value of the input signal is $P_{smx} = A_{mx}^2/2$. Combining (1.5) and (1.11), we have:

$$\rho_{mx}^2 = \frac{1}{8\pi^2 K_\delta} \times \frac{f_s^3}{Bf^2} \tag{1.12}$$

It is possible to conclude from the derived eqs. that the LΔM signal-to-quantization noise ratio is dependent on the level of input signal and the maximum value of this ratio is inversely proportional to the square of input frequency. Strong dependency of this ratio on the input amplitude level as well as the input frequency is a serious disadvantage of LΔM in comparison to PCM and Δ–ΣM. It is necessary to point out that derived relations hold only for systems without slope overload, and with uniformly distributed quantization noise.

To find the amplitude dynamic range D for LΔM system we have to find the minimum value of the input signal amplitude that LΔM is able to convert correctly. From Fig. 1.2, it is evident that $A_{min} > \delta/2$, amplitude dynamic range is defined as:

$$D = \frac{A_{mx}}{A_{\min}} = \frac{1}{\pi}\left(\frac{f_s}{f}\right) \tag{1.13}$$

The dynamic range can be seen to be dependent on the ratio of $f_s$ to $f$. In the case of very small amplitude of input signal, the output of LΔM will be a rectangular pulse stream of amplitude $\delta/2$ with a period of $2T_s$. The mean power of the signal in this extreme case is $\delta^2/4$. In normal working conditions (no slope overload) sampling frequency, $f_s = 1/T_s$, is much higher than the cut-off frequency of the input signal, and because of that the first sub-harmonic frequency of the clock $(f_s/2)$ does not pass through the band pass of the receiving filter of the demodulator.

In conclusion, we can say that a linear delta modulator is the first order differential pulse code modulation system (DPCM). It shapes the spectrum of the modulated signal. The modulated signal presents the first order difference between the input and the reconstructed signal. In fact, it presents the error. This is the reason why these types of converters are known as error encoders as well. At the receiving (demodulation) side, the quantization noise stays unchanged. To restore the signal, a low-pass filter is needed.

In addition to the slope overload problem, the LΔM system is sensitive to channel errors in transmission because the receiver consists of a memory element (integrator). This can be very critical when LΔM is used for conversion of low-level telemetry signals. Fig. 1.4 shows the case when the transmitting signal is a low level dc signal in the presence of channel error. It can be seen that the reconstructed signal at the receiver describes quite a different signal level.

In spite of its disadvantage LΔM can be used in video and audio applications, PCM conversion and instrumentation [2].

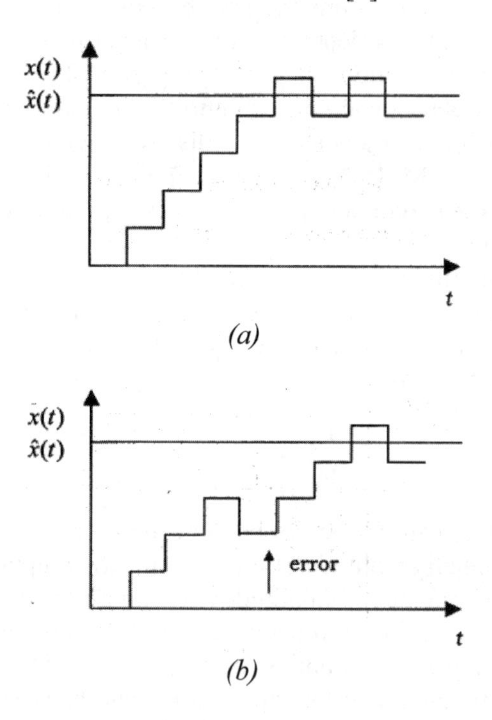


**Fig. 1.4. (a)** Receiver waveform, no error in transmission, **(b)** receiver waveform when a single error occurs

## 1.2 DELTA-SIGMA MODULATION SYSTEMS

As stated earlier, LΔM has several disadvantages such as the accumulation of error in the integrating receiver. This is a serious problem when the dc signal has to be transmitted. Its amplitude dynamic and signal-to-quantization noise ratio are inversely proportional to the frequency of the input signal. All of these disadvantages of the LΔM system stem from the fact that the output of the LΔM encoder is the result of the equivalent differentiation of a continuous input signal. As mentioned earlier, ordinary LΔM is suitable for encoding signals whose spectral power density is decreasing with the increase of the signal frequency to be encoded. This is true for speech signals. Video signals and different telemetry signals have more or less uniform spectra. For such signals, it is suggested to use delta-sigma modulation [6]. We can find in the literature the name of sigma-delta modulation as well [7]. Both names are in use, but we prefer to use the name given in the original papers of the inventors [6].

### 1.2.1 The Principle of Δ–ΣM

The basic idea of Δ–Σ modulation consists in adding an integrator in front of the ordinary LΔM, and a differentiator in front of the linear delta demodulator. As can be seen from fig. 1.5, Δ–ΣM requires two integrators, thus the difference signal is

$$\varepsilon(t) = m(t) - \hat{x}(t) \tag{1.14}$$

or

$$\varepsilon(t) = \int_0^t \left(x(t) - \delta Y_n\right) dt \,. \tag{1.15}$$

Since subtraction is a linear operation, and because an integrator is a linear operator, it does not matter if we do the summation first and then integration. Because of the cancellation of operations, differentiation and integration, the delta-sigma demodulator becomes a low pass filter (an averager). Fig. 1.5 becomes the arrangement shown in Fig. 1.6.

This structure is simpler and can be considered as a “smoothened version” of a linear delta modulator. Sometimes, the Δ–Σ modulator is referred to as a noise shaping or an interpolative encoder. For now we will mention that noise shaping is dependent on the order of summation circuit (Σ), and that channel errors are now smoothened by an averaging filter. This simple, but ingenious idea of Japanese inventors opened a wide range of applications [6], [16], thanks to the latest advances of VLSI technology. It took almost twenty years to recognize that the differentiation property of the LΔM system imposes a serious problem at the LΔM receiver. The differentiation can be considered as a micro change (local) and the smallest change in the channel can have serious changes on the received signal. From another point of view, we can consider integration to be a macro change and small changes do not have a significant influence on the received signal.

To understand the principle of operation of Δ–ΣM, let us consider an idle pattern generated in Δ–ΣM. This pattern depends on the order of the noise-shaping filter in the encoder (a modulator's order indicates the number of integrators or the order of the analog filter in the loop).

Throughout this book we will be limited to the first order Δ–ΣM only. We will use the intuitive approach shown in Table 1.1 to demonstrate operation of a delta-sigma modulator. A first order modulator with labeled nodes is shown in Fig. 1.6. Suppose the full-scale swing of the binary quantizer (BQ) is $\pm 1V$.

At the summing amplifier (delta), the $+1V$ or $-1V$ is subtracted from the analog input voltage. The signal of difference $\varepsilon(t)$ is the input to the integrator (sigma). The input voltage to the integrator (accumulator) is added to the old value of the accumulator, and the voltage at the node $s(t)$ becomes the new voltage on the node $s(t+\tau)$. The voltage $s(t)$ is compared in the binary quantizer to the ground:

If $s(t) \geq 0$, then $Y_n = +1V$;

If $s(t) < 0$, then $Y_n = -1V$;

Each operation occurs once during each clock cycle. In the example shown in table 1.1, analog input voltage, $x(t) = 0.8V$, and all initial values of node voltages are set to zero.

As can be seen from table 1.1, all node voltages are identical in clock periods two and twelve. If the analog input stays the same, the same pattern will repeat itself. Thus the average value of the demodulated output $y(t) = 8/10 = 0.8V$ yields a numerical representation of the analog input.

From Table 1.1, we can see some stable patterns for dc input signals of a first-order Δ–ΣM. The drawback of stable patterns is found in the frequency distribution of the quantization error [5]. For the purpose of this book, the quantization error is considered to have a uniform distribution, and we will accept the "linear model" of Δ–ΣM. Unfortunately, the linear model breaks down for certain types of input signal of a first-order Δ–ΣM. When Δ–ΣM exhibits a stable pattern, correlations with the input signal are obtained. This can be a problem in audio applications, because the human ear is very sensitive to repetitive signals. In Fig. 1.7, the signal of difference $\varepsilon(t)$ and its integrated version are shown. From Fig. 1.8, we can see that for positive peak amplitude of input, the output of the modulator stays high, a logic one most of the time. When the sine wave is moving through the middle value, the output bounces back and forth. When the input signal is approaching the negative peak amplitude, the modulator stays low, logic zero most of the time.

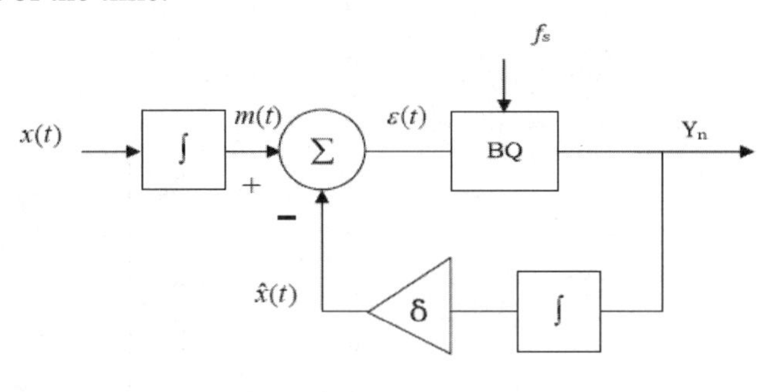


*(a)*

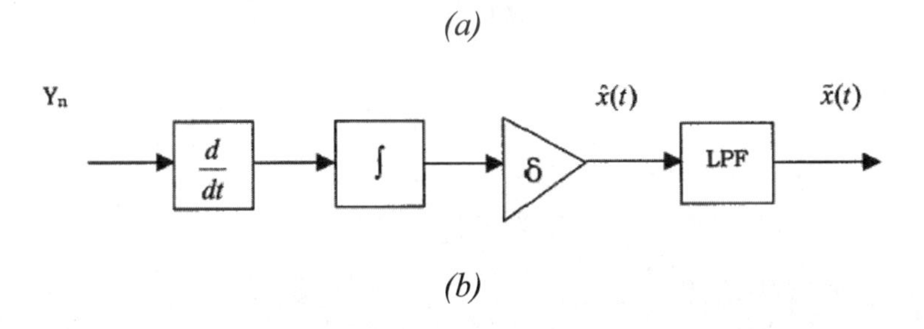


*(b)*

**Fig. 1.5.** Block diagram of a Δ-ΣM system, **(a)** modulator, **(b)** demodulator

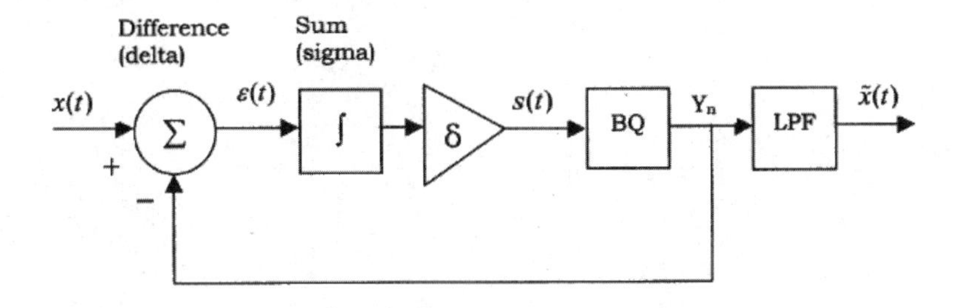


**Fig. 1.6.** Block diagram of a simplified Δ-ΣM system

**Table 1.1.**

| Clock period | $\varepsilon(t)$ | $s(t)$ | $Y_n$ | Average over period |
|---|---|---|---|---|
| 0 | 0 | 0 | 0 | |
| 1 | 0.8 | 0.8 | 1 | |
| 2 | -0.2 | 0.6 | 1 | |
| 3 | -0.2 | 0.4 | 1 | |
| 4 | -0.2 | -0.2 | -1 | |
| 5 | -0.2 | 0 | 1 | |
| 6 | -0.2 | -0.2 | 1 | |
| 7 | -1.8 | 1.6 | 1 | → = 0.8V |
| 8 | -0.2 | 1.4 | 1 | |
| 9 | -0.2 | 1.2 | 1 | |
| 10 | -0.2 | 1.0 | 1 | |
| 11 | -0.2 | 0.8 | 1 | |
| 12 | -0.2 | 0.6 | 1 | |

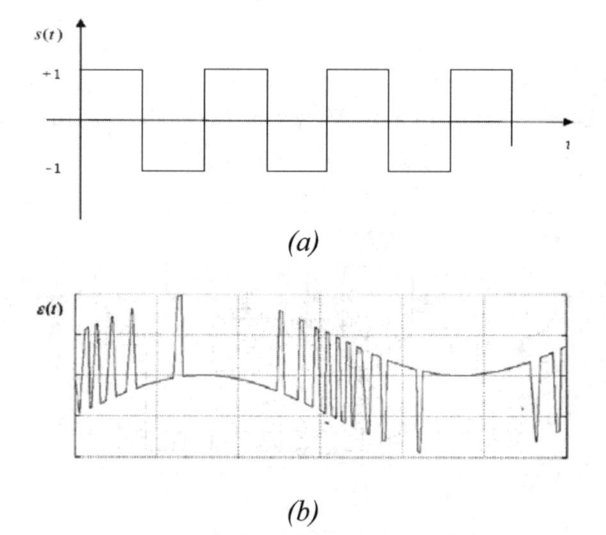


*(a)*

*(b)*

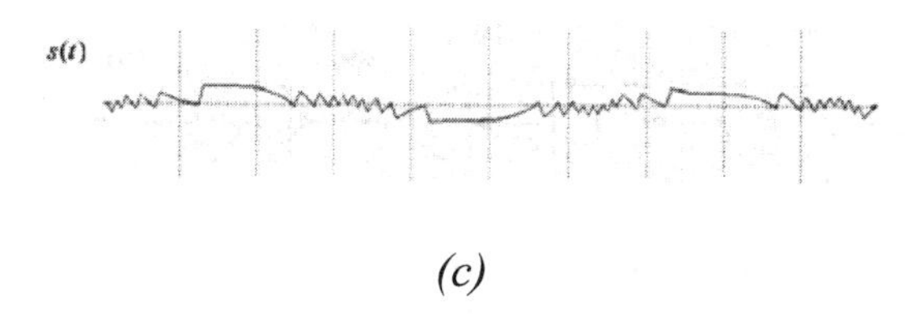


(c)

**Fig. 1.7.** Idle pattern of Δ-ΣM, **(a)** zero input, **(b)** difference signal, **(c)** output of a 1st order integrator

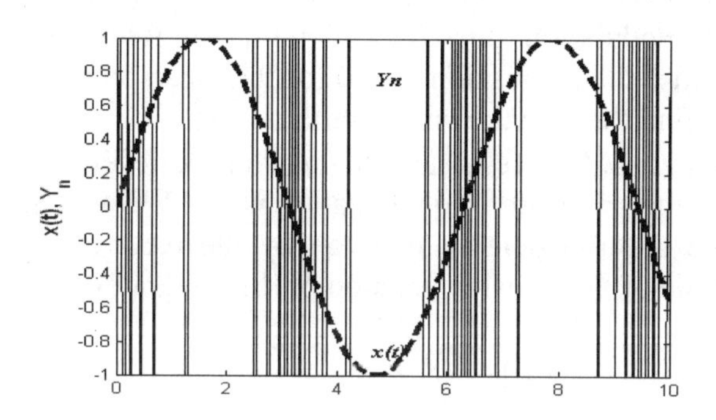


**Fig. 1.8.** Input and output pattern of Δ-ΣM

### 1.2.2 Basic Parameters of Δ–ΣM

For mathematical analysis of Δ–ΣM, it is useful to consider the block diagram in Fig. 1.5. Suppose that the input signal $x(t)$ is a sine of amplitude $A$. Let

$$x(t) = A\cos 2\pi ft \,. \tag{1.16}$$

As shown in Fig. 1.5, the input signal passes through an integrator first. The output of this filter is

$$\int x(t)dt = \frac{A}{2\pi f}\sin 2\pi ft \,. \tag{1.17}$$

We suppose that the integration constant is equal to zero. The maximum slope of this signal is

$$\left|\frac{d}{dt}\left(\frac{A}{2\pi f}\sin 2\pi ft\right)\right|_{t=0} = A \cdot \tag{1.18}$$

On the other hand, for the first-order Δ–ΣM, the maximum slope is $\delta/T_s$. Thus to avoid slope overload, the following relation must be satisfied,

$$A_{mx} < \frac{\delta}{T_s} = \delta f_s \cdot \tag{1.19}$$

We can see from (1.19) that the maximum allowed amplitude of the input signal, for Δ–Σ modulation, does not depend on the frequency of the input signal. We can use the expression (1.10) for the mean power of the quantization noise to find the signal-to-noise ratio.

However, for Δ–ΣM there is no integrator at the receiver terminal. We need to modify the spectrum of quantization noise. We accomplish this by multiplying the spectral density function by the reciprocal value of the square magnitude of the transfer function of the integrator,

$$\Phi(f) = \frac{\rho_\delta}{B} = K\frac{\delta^2}{f_s} \cdot \tag{1.20}$$

For Δ–ΣM, the spectral power density of quantization noise at the input of the receiver (demodulator) is

$$\Phi(f) = K\frac{\delta^2}{f_s}(2\pi f)^2 \cdot \tag{1.21}$$

because the transfer function of an ideal integrator is $H(j\omega) = 1/j\omega$. The power of quantization noise at the output of an ideal low-pass filter (LPF) of bandwidth B [Hz] is

$$P_{\Delta-\Sigma} = 4\pi^2 K\frac{\delta^2}{f_s}\int_0^B f^2 df = \frac{4}{3}K(\delta)^2\frac{B^3}{f_s} \cdot \tag{1.22}$$

For the sinusoidal input signal of amplitude $A$, a mean power signal-to-quantization noise ratio is

$$\rho^2_{\Delta-\Sigma} = \frac{P_s}{P} = \frac{3}{8\pi^2 K}\left(\frac{f_s}{B^3}\right)\left(\frac{A}{\delta}\right)^2 \cdot \tag{1.23}$$

The maximum value of this ratio is

$$\rho^2\Big|_{A=\frac{\delta}{T_s}} = \frac{3}{8\pi^2 K}\left(\frac{f_s}{B}\right)^3. \tag{1.24}$$

This expression is similar to (1.12) for maximum signal-to-quantization noise ratio of ordinary LΔM. In both expressions the signal-to-quantization noise ratio is directly proportional to the cube of the sampling frequency $f_s$. However, it is very important to note that in the case of Δ–ΣM, this ratio does not depend on the frequency of the input signal.

It is important to point out once again that all the derived relations hold only for a uniform distribution of quantization noise, i.e. a linear model of Δ–ΣM. To accurately modulate the signal, the minimum value of the amplitude of the analog input has to be slightly higher than $\delta$. This means that the threshold value of the input signal equals $\delta/2$. When a small input signal is applied to the modulator, the sum (or integral) of the input samples adds to the quantizer input. The sensitivity of the idle pattern depends on the initial state of the filter. It is the worst for initial values $X_0 = 0$ and $X_0 = \delta/2$. In the case that the input signal is a sine wave, $x(t) = A\sin 2\pi ft$, and the sum of the input samples can be approximated by,

$$\sum_K x(kt) \approx \int_0^{KT_s} A\sin 2\pi ft dt = -\frac{A}{2\pi f}\left[\cos 2\pi fkT_s - 1\right] \tag{1.25}$$

The sum of input samples has an absolute maximum

$$\left|\sum_K x(kt)\right|_{\max} \approx \frac{A}{\pi f}$$

In the case that this absolute maximum exceeds the value of $\delta/2$, the minimum value of the input signal has to be, $A_{min}/\pi f > \delta$ or $A_{min} > \delta\pi f$. The amplitude dynamic range of Δ–ΣM is

$$D_{\Delta-\Sigma} = \frac{A_{\max}}{A_{\min}} = \frac{1}{\pi}\left(\frac{f_s}{f}\right). \tag{1.26}$$

This expression is identical to expression (1.13). This means that the amplitude dynamics of ordinary LΔM and Δ–ΣM are identical, in spite of the fact that expressions for minimum and maximum amplitude are different. With some disappointment we can conclude that even for Δ–ΣM, amplitude dynamic decreases with frequency increase of the input signal.

### 1.2.3 Linear model of Δ–ΣM

The binary quantizer (BQ) in the delta-sigma modulator loop presents a 1-bit ADC. An uncertainty of any ADC, or quantization error, is equal up to ±1/2 < LSB (least significant bit). In the case of Δ–ΣM, this error is $\pm\delta/2$. Even though this error can be correlated for certain types of input signal [5], we will assume that this error "signal" is totally random (uncorrelated with the input). This can be achieved with a proper design [7], and we will assume to be white noise, with its energy spread uniformly over the band from dc to one-half the sampling rate. The RMS value of noise source relative to the input can be shown to be $(6.02N + 1.76)$ dB for an N-bit resolution converter. Because Δ–ΣM is 1-bit ADC, it offers an almost comical 7.78 dB signal-to-noise ratio. However, the input signal is grossly oversampled [8], [9]. For example, if the sampling frequency is 3MHz, then the quantization noise is spread over a wide range of 1.5 MHz. If the bandwidth of interest of input signal is speech signal of 4 kHz, then the noise density is reduced. In addition, the high order analog filters are used in the modulator loop to further reduce noise density in the band of interest. This filter shapes the quantization noise spectrum, and that is the reason why this type of ADC is named by some authors as a noise shaping converter. The noise-shaping principle is illustrated by using a simplified frequency or $s$-domain model. In Fig. 1.9, a linearized model of a first-order Δ–Σ modulator is shown.

The comparator acts in a linear model as a noise source. Using the principle of superposition we can write: First, $N(s) = 0$ then the so-called signal transfer function (STF) is:

$$Y_1(s) = [X(s) - Y_1(s)]\frac{1}{s} \tag{1.27}$$

or

$$\frac{Y_1(s)}{X(s)} = \frac{1}{s+1} \tag{1.28}$$

This is the low-pass filter and its characteristic is sketched in Fig. 1.10. Second, $X(s) = 0$, then so-called noise transfer function (NTF) is:

$$Y_2(s) = -Y_2(s)\frac{1}{s} + N(s) \tag{1.29}$$

$$\frac{Y_2(s)}{N(s)} = \frac{s}{s+1}. \tag{1.30}$$

This is the transfer function of a high-pass filter and its frequency response is shown in Fig. 1.11.

From fig. 1.9, we can conclude that as the loop integrates the error, the difference $X(s)$ - $Y(s)$, it low-pass filters the signal and it high-pass filters the noise. The $\Delta$–$\Sigma$ loop pushes the quantization noise into a higher frequency band. The input signal is left unchanged as long as its frequency content doesn't exceed the filter's cutoff frequency. Combining (1.28) and (1.30) we have:

$$Y(s) = \frac{X(s)}{s+1} + \frac{s}{s+1} N(s) \tag{1.31}$$

Note that at $s = 0$ (a frequency of $f = 0$), the output equals $X$ with no quantization noise. At higher frequencies, the value of $x$ is reduced and the value of quantization noise is increased. At frequency $f = \infty$, the output equals only noise. The analog filter has a low-pass effect on the signal and high-pass effect on the noise. A sketch of power spectral density of a first order delta-sigma modulator is shown in Fig. 1.12. It is evident that with the increase of order of the filter $H(s)$, the quantization noise can be pushed into a higher frequency range.

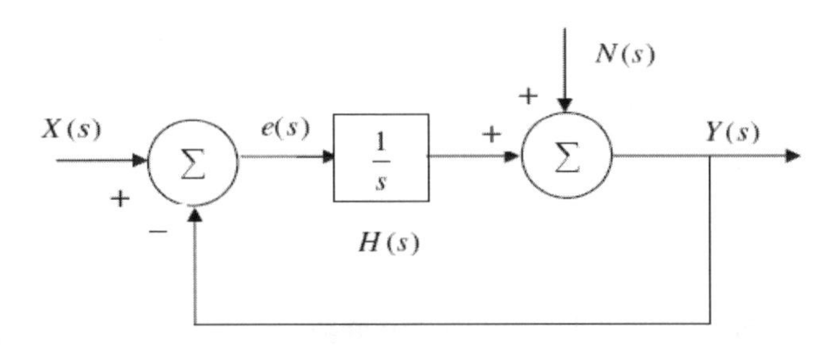


**Fig. 1.9.** Linearized model of Δ-ΣM

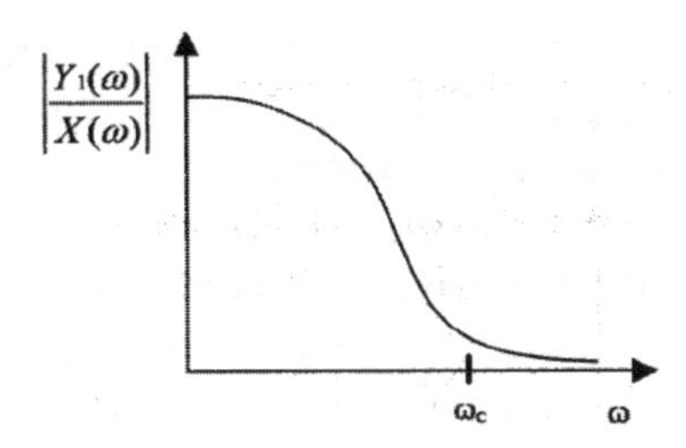


**Fig. 1.10.** Amplitude sketch of STF

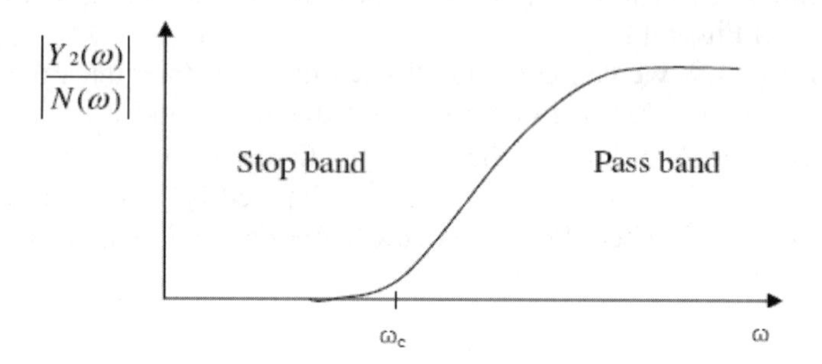


**Fig. 1.11.** Amplitude sketch of NTF

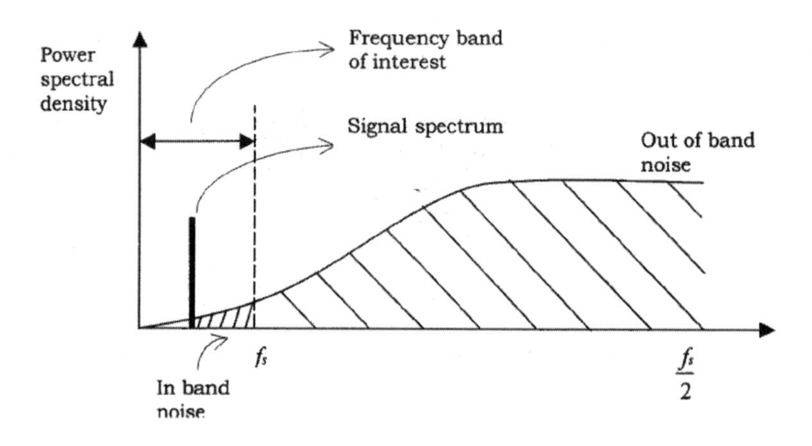


**Fig. 1.12.** Noise shaping in the frequency domain

### 1.2.4 Anti-Aliasing Requirements

It is well known that the input spectrum of any ADC repeats around integer multiples of its sampling rate, Fig. 1.13. A delta-sigma ADC does not provide noise rejection at the region around integer multiples of the sampling rate ($nf_s$). Since delta-sigma ADCs are grossly over-sampled, anti-alias filtering (AAF) is often trivial. Often a single pole, passive RC filter at the input of Δ–ΣM is sufficient in most applications. This is a big advantage in comparison to pulse-code modulation (PCM) ADC.

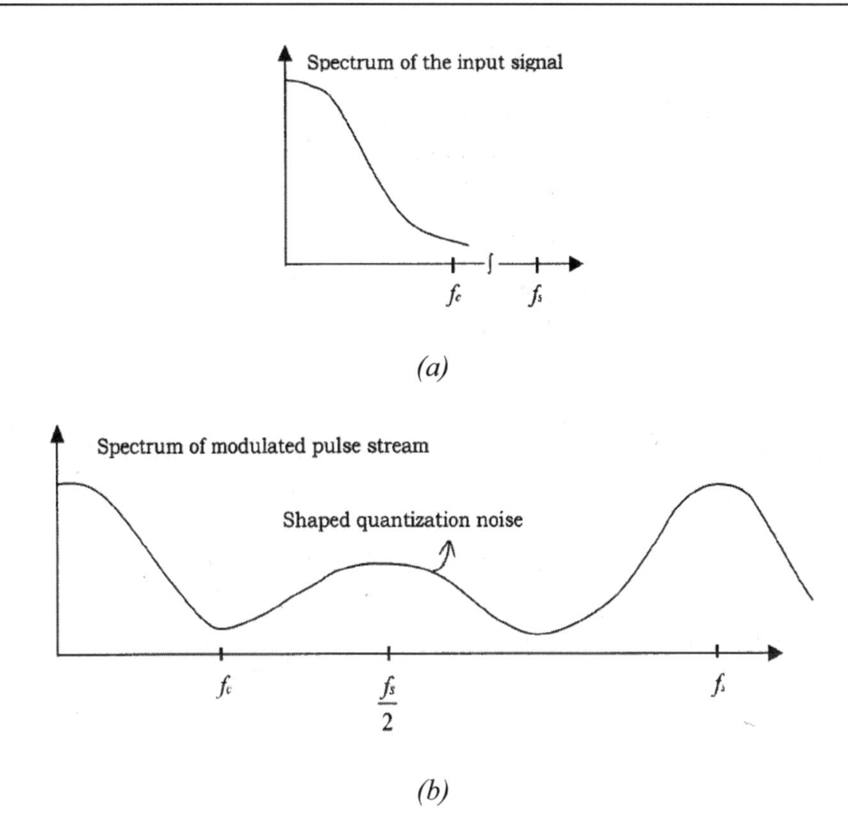


**Fig. 1.13.** Sketch of amplitude spectrum, (a) input and (b) Δ-ΣM pulse stream

## 1.3 MULTI-LEVEL DELTA MODULATION SYSTEMS

A pulse code modulation (PCM) is a classic example of multi-level ADC. These systems are open loops, having quantizers with many levels. A typical PCM encoder for speech applications has 256 quantization levels and uses a companding technique to achieve adaptive quantization. While PCM systems are open loop systems without memory having quantizers with many levels, ΔM systems are closed loop systems with memory having a quantizer embedded in the forward path. ΔM systems with a binary quantizer are discussed in the previous section. In this section, we will discuss the ternary delta modulation system only.

### 1.3.1 Signal-to-Noise Ratio

Replacing BQ in the LΔM encoder, shown in Fig. 1.1, with a quantizer having more than two levels, the resulting system is called multi-level ΔM [2], Fig. 1.14.

If the transmission is restricted to the binary type, then $X_n$ samples are binary encoded at a rate $n\,f_s$, where $n$ is the number of bits required to encode each of the $N$ quantization levels and $fs$ is the clock rate of the multi-level ΔM encoder. The binary transmitted waveform $X_n$ must be decoded at the receiver prior to integration and low-pass filtering [2]. Fig. 1.14 shows the arrangement.

The theory of optimum quantizer is given in [2]. Here, we will give a short interpretation of signal-to-quantization noise ratio (SQNR) for multi-level quantizer. According to Still [2],

$$SQNR = \frac{\sigma_x^2}{D} = \frac{N^k}{d}\left(\frac{\sigma_x}{\sigma_e}\right)^2 \tag{1.32}$$

where $\sigma_x^2$ is a mean square value of input signal $x(t)$, $\sigma_e^2$ is a mean square value of error signal $\varepsilon(t)$, N number of quantization levels, $D = dN^{-k}\,\sigma_e^2$, $d$ and $k$ are constants whose values depend on the number of quantization levels $N$. When $N$ is small, say 4 to 8 levels, $d \approx 1.3$, and $k \approx 1.76$. The SQNR becomes [2]:

$$SQNR = 5.3n - 1.14 + 10\log_{10}\left(\frac{\sigma_x}{\sigma_e}\right)^2 \; [dB] \tag{1.33}$$

These formulas are derived under the condition when the encoder is correctly tracking a Gaussian input signal $x(t)$. Here again we consider the linear model of a multi-level quantizer, and spectral density fraction of the quantization error is relatively flat over the message band. This spectral density function has a peak of $f_s/2$ and a minimum in the region $f_s/4$ when a slope overload does not occur.

### 1.3.2 Ternary ΔM System (TΔM)

A theory derived for LΔM holds for TΔM as well. A difference exists in the quantizer only. The decoder is identical to the decoder of LΔM. Replacing the binary quantizer BQ with the quantizer of three levels, ternary ΔM is achieved. Multi-level ΔM is in fact a generalization of LΔM with BQ. The reconstruction signal of the ternary decoder is changing for the amount $-\delta$, 0, $+\delta$, depending on the ratio of input and reconstruction signal.

In fig. 1.15 the arrangement of the ternary ΔM system is shown. The signal of difference $\varepsilon(t)=x(t)-\hat{x}(t)$ is connected to the input of two comparators. The operation of TΔM is as follows:

$$X_n = \begin{cases} 1, & \frac{\delta}{2} < \varepsilon(t) \\ 0, & -\frac{\delta}{2} \le \varepsilon(t) \le \frac{\delta}{2} \\ -1, & \frac{\delta}{2} > \varepsilon(t) \end{cases} \tag{1.34}$$

If $X_n = 1$, this means that $x(t)$ is greater than $\hat{x}(t)$. The ideal integrator will be charged for the amount of $\delta$ if $X_n = -1$, the ideal integrator will be discharged for the same amount of $-\delta$. If $X_n = 0$, the ideal integrator will keep its previous value. For the same step size $\delta$ and the same sampling frequency TΔM, the average number of pulses (+1 or -1) is significantly lower in comparison to LΔM. In addition to better signal-to-quantization noise ratio, TΔM is closer to the natural choice. Fig. 1.16 shows relevant waveforms of TΔM when the input is a sinusoidal signal. There are a number of applications where the third stage is needed. For example, count up down, and stop; shift left right and stop, etc.

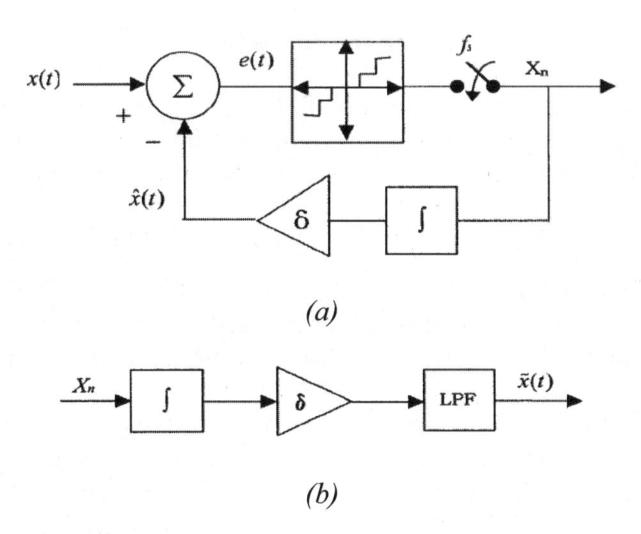


**Fig. 1.14.** Multi-level ΔM system, **(a)** modulator, **(b)** demodulator

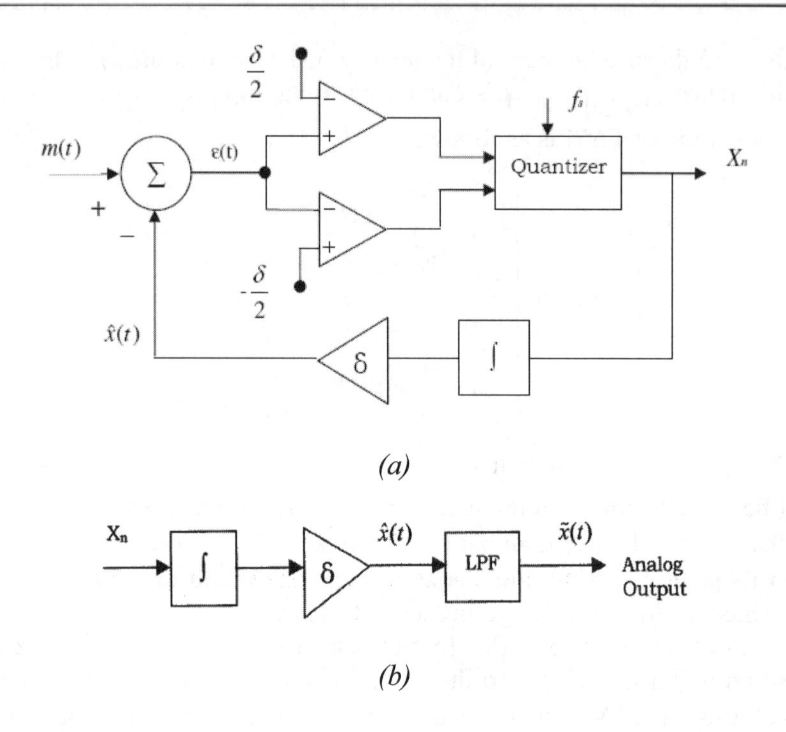


**Fig. 1.15.** Ternary ΔM systems, **(a)** modulator, **(b)** demodulator

### 1.3.3 Tri-level Delta-Sigma Modulation

Adding an integrator at the input of TΔM, and adding a differentiator at the input of demodulator in fig. 1.15, a ternary delta-sigma modulation system results. Similarly, as in the case of binary Δ-ΣM, we can write

$$\varepsilon(t) = m(t) - \hat{x}(t), \text{ or } \varepsilon(t) = \int x(t)dt - \int X_n dt = \int \left(x(t) - X_n\right)dt \qquad (1.35)$$

Since subtraction is a linear operation, we can transform fig. 1.15 into fig. 1.17, where ternary ΔM becomes ternary delta-sigma modulator. Demodulator became LPF because operations of differentiation and integration cancel.

As in the previous case, two comparator thresholds are used to establish a dead zone around zero. Again, if the accumulated error is large and positive, i.e. $\varepsilon(t) > \delta/2$, a code of +1 is produced. If the accumulator error is large and negative, i.e. $\varepsilon(t) < -\delta/2$ a code of -1 is produced. However, if the accumulated error is small, i.e. $-\delta/2 \le \varepsilon(t) \le \delta/2$, a zero-level signal is produced. Fig. 1.18 shows the output of the integrator $\varepsilon(t)$ and the ternary

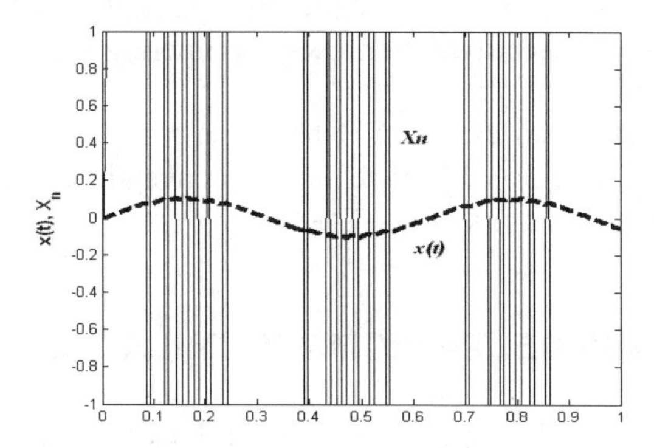


**Fig. 1.16.** Reconstruction signal $\hat{x}(t)$ and ternary output $X_n$

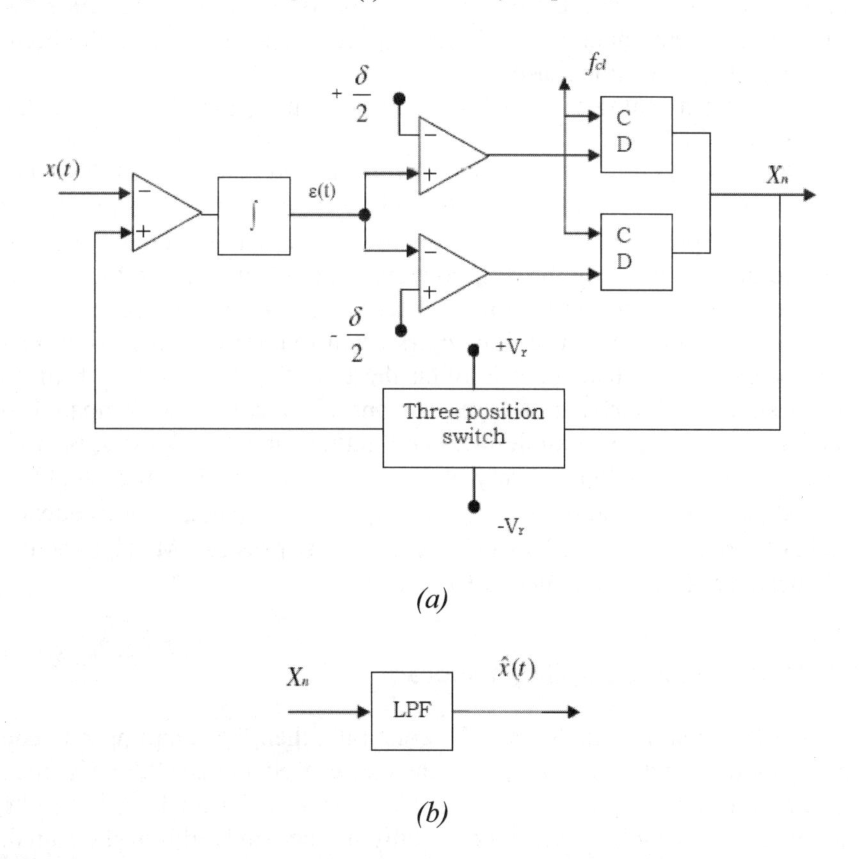


**Fig. 1.17. (a)** Ternary delta-sigma modulator, **(b)** demodulator

modulated sequence $X_n$. Fig. 1.19 shows the sinusoidal input and demodulated output. We can see that after low-pass filtering (the third-order Butterworth filter) output is phase shifted.

The simulation of tri-level Δ–ΣM has been reported in reference [10]. Improvement of more than 15 dB relative to conventional LΔM coding has been demonstrated for a first order TΔ–ΣM.

## 1.4 BAND-PASS DELTA-SIGMA MODULATION

With growing demand for portable wireless devices, recent efforts in the design of the integrated circuits for radio frequency (RF) communication receivers have focused on band-pass delta-sigma modulation. Oversampled delta-sigma modulators are uniquely suited to this application because the adjacent channel interferers fall into the same band as the high-pass shaped quantization noise.

Shifting from analog to digital signal processing generally increases the burden on the delta converters that provide the interfaces between the analog and digital circuits. For example, if it is desired that much of the channel filtering in the receiver be performed by digital filter, then the digital filter must be preceded by ADCs with sufficient dynamic range and bandwidth to digitize not only the desired signal but also the interfering signals to be removed by the digital filters. This creates a potential problem because high performance data converters often require high-precision analog processing. Fortunately, it is often the case that the bandwidth of an analog signal of interest in a wireless transceiver is narrow compared to practical data converter sample-rates and digital filters clock rates, so high analog precision is only necessary within the narrow band of interest [11].

Band-pass Δ–Σ modulators suppress quantization noise in frequency band not centered at dc [13], as in the case of low-pass Δ–ΣM. This section is dedicated to basic principles of BPΔ–ΣM.

### 1.4.1 Band-pass Sampling Theorem

If the IF signal is sampled at a Nyquist rate, then the sampling rate can be ridiculously high. For example, if the carrier frequency is 2.5 GHz, then the sampling rate required can be at least 5 GHz. Fortunately, it can be shown that the sampling rate depends only on the bandwidth of the signal, not on the absolute frequency involved. This is equivalent to saying that we can reproduce the signal from the samples of the complex envelope [14].

*Band-pass Sampling Theorem:*

If a real band-pass waveform has a nonzero spectrum only over the frequency interval $f_{c1} < |f| < f_{c2}$, when absolute bandwidth $B = f_{c2} - f_{c1}$, then the waveform may be reproduced from sample values if the sampling rate is

$$f_s \geq 2B\,. \tag{1.36}$$

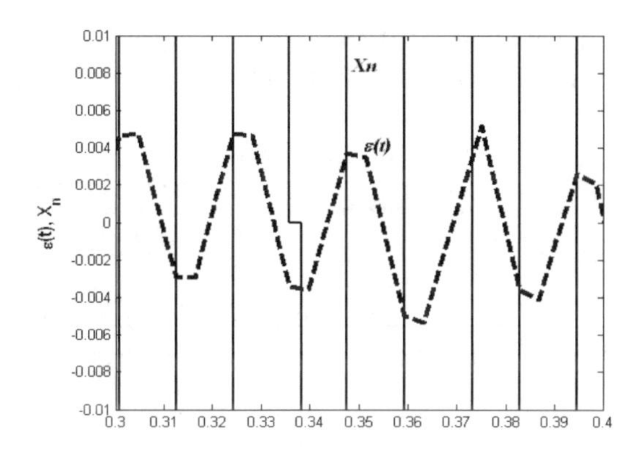


**Fig. 1.18.** Relevant waveforms of T-ΔM

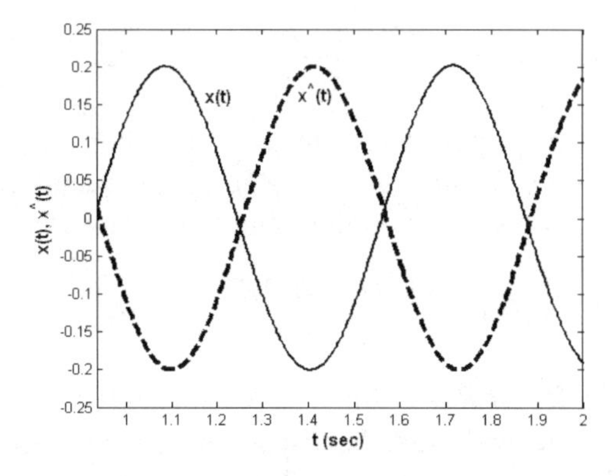


**Fig. 1.19.** Sinusoidal input $x(t)$ and demodulated output $\hat{x}(t)$

A basic premise of delta-sigma modulation is that the sampling rate is much greater than the highest frequency of interest present in the input. This is necessary because low-pass delta-sigma modulators have zero quantization noise only near dc. If one were to nullify quantization noise at some other frequency, say $\omega_0$, then one would obtain good accuracy in a band centered on $\omega_0$. Therefore, BPΔ–ΣM are useful in situations where an analog input signal of interest is not centered at dc, rather than down-converting the real and imaginary components of the desired signal to dc and digitizing the resulting two signals with a pair of low-pass Δ–Σ modulators. A single band-pass Δ–Σ modulator can be used instead as in [11], [13].

The formula for calculation of the signal-to-quantization noise ratio is derived by Sehreier and Snelgrove [13]. The error transfer function of order $n$ is,

$$|H(e^{jw})| = K(\omega - \omega_0)^n, \text{ where } K = \frac{1}{n!} \frac{d^n |H(e^{jw})|}{d\omega}\Bigg|_{\omega=\omega_0} \tag{1.37}$$

If we assume that the quantization error is white noise and uniformly distributed in the range [-1, +1], then the one-sided noise power is given by:

$$N_q^2 = \int_{\omega_0}^{\omega_0+\omega_B} |H(e^{jwt})| \frac{e^2}{\pi} d\omega = \frac{K^2 \omega_N^{2n+1}}{3\pi(2n+1)} \tag{1.38}$$

If an input signal power of sinusoidal input is $A^2/2$, then the signal-to-quantization noise ratio of an $n^{th}$ order band-pass converter is:

$$SNR = 10 \log \frac{3A^2(n+1)(2R)^{2n+1}}{4K^2\pi^{2n}} \quad dB \tag{1.39}$$

The pass band of a band-pass converter is $B$, and then we have to integrate noise from $(\omega_0 - \omega_B/2)$ to $(\omega_0 + \omega_B/2)$. Consequently, the signal-to-quantization noise ratio of an $n^{th}$ order band-pass converter is:

$$SNR = 10 \log \frac{3A^2(n+1)(2R)^{n+1}}{4K^2\pi^{n}} \quad dB \tag{1.40}$$

The over-sampling factor $R$ is defined as $R = \pi/\omega_B$ [13].

The tunable, resonator-based, band-pass Δ–Σ modulator was proposed in reference [15]. It was found that inclusion of a double delayer (see reference [16], for example) in the feedback is not sufficient. To improve signal-to-quantization noise ratio (SNR), the author of reference [15] proposed the inclusion of a single delayer whose associated coefficient

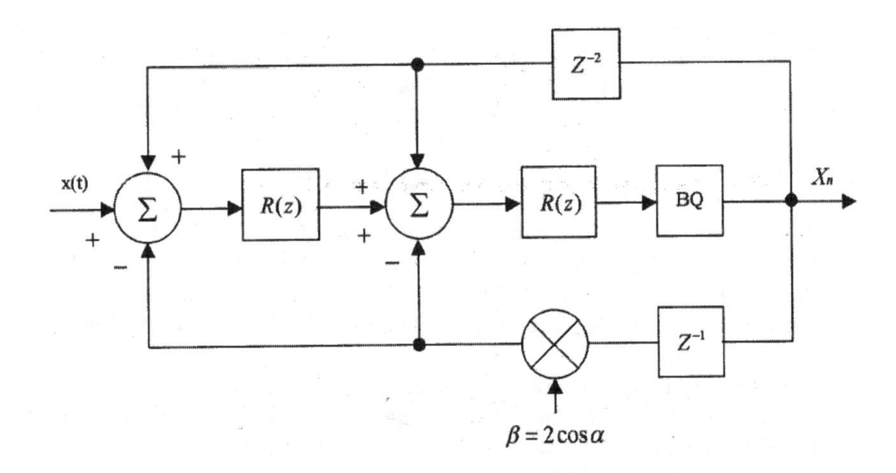


**Fig. 1.20.** Block diagram of tunable center frequency *BP*Δ-ΣM [15]

depends on $\omega_0$. The block diagram of a second order variable center frequency BPΔ-Σ modulator is shown in Fig. 1.20. For the linearized system of Fig. 1.18, the signal transfer function is given by:

$$H_s(z) = \frac{R^2(z)}{1 - R^2(z)F(z) - R(z)F(z)} \tag{1.41}$$

and the noise transfer function:

$$F_N(z) = \frac{1}{1 - R^2F(z) - R(z)F(z)} \tag{1.42}$$

where $F(z)$ represents compensation hardware in the feedback path, and $R(z)$ is the variable centre frequency resonator transfer function. Criteria for design and stabilization of $H_s(z)$ and $F_n(z)$ are given in reference [15] and the reported simulation results show a 5dB improvement in SNR over 80dB dynamic range.

## 1.5 CONCLUSION

In this chapter, the basic concept of over-sampling ADC has been introduced. We presented several architectures such as LΔM, Δ-ΣM, TΔM, and BPΔ-ΣM. The basic parameters for the linear models of LΔM and Δ-ΣM were derived. Additionally operation of tri-level Δ-ΣM and BPΔ-ΣM were described.

## REFERENCES

1. Al-Janabi M., Kole F., and Morling R. C. S., "Variable Centre Frequency Resonator-Based Bandpass ΣΔ Modulator" Electronic Letters.
2. Analog Devices: www.analog.com/sigma-deltaADCs.
3. Candy J., Temes G., Editors, Oversampling Delta-Sigma Data Converters, Theory, Design and Simulation, IEEE Press 1992.
4. Couch L.W., Digital and Analog Communication Systems, Sixth Edition, Prentice Hall, 2001, ISBN: 0-13-08-1223-4.
5. Crystal Semiconductor Corporation, Data Book Vol. 1, Analog/Digital IC's, Austin, Texas, 1990.
6. Deloraine E.M., Van Miero S., and Derjavich B., French Patent #932140.
7. Feldman A., Boser B., Gray P., "A 13-bit Sigma-Delta Modulator for RF Baseband Channel Applications", IEEE Journal of Solid-State Circuits, Vol. 33, No. 10, October 1998.
8. Galton I., "Delta-Sigma Data Conversion in Wireless Transceivers" IEEE Tr. on Microwave Technology and Techniques, Vol. 50, No. 1, January 2002, pp 302-315.
9. Inose H., Yasude Y., and Murakami J, "A Telemetry system by Code Modulation - ΔΣ Modulation", IRE Trans. Space Electron. Telemetry, vol SET8, pp 204-209, Sept 1962.
10. Lukatela G., Drajic D., Petrovic G., Digital Communications, Nolit, Beograd, 1978.
11. Motorola, In Catalog on Telecommunications, CVDSM MC3417.
12. Norsworthy S., Schreier R., and Temes G., Editors, Delta-Sigma Data Converters, IEEE Press, New York, 1997, ISBN: 0-7803-1045.
13. Paulos J., Brauns G., Steer M, Ardolan S., "Improved signal-to-noise ratio using tri-level Delta-Sigma Modulation", IEEE Conference on CAS, 1987, pp. 463-466.
14. R.M. Gray, "Oversampled Sigma-Delta Modulation", IEEE Trms. On Commun., Vol. 35 No. 5, pp 481-489, May 1987.
15. Schreier R., Snelgrove, M., "Bandpass Sigma-Delta Modulation" Electronic Letters, 9th November 1989, Vol. 25, No. 23, pp. 1560-1561.
16. Still R., Delta Modulation Systems, Pentech Press, London, 1975, ISBN: 0-7273-0401-1.
17. Van Englen J., Van de Plassehe R., Bandpass Sigma-Delta Modulators, Kluwer Academic Publishers, Boston, 1999, ISBN: 0-7923-8698-1.

# CHAPTER 2 LINEAR ARITHMETIC OPERATIONS

## 2.1 INTRODUCTION

In general, the goal of this chapter is to introduce a "compromise" between amplitude and time quantization. This compromise exists in all digital systems. The accuracy of time quantization is much easier to implement than accuracy of amplitude quantization. For example, the stability of sampling frequency of $10^{-8}$ or greater is easy to achieve using crystal oscillators, while the stability of reference amplitude levels of $10^{-3}$ presents a serious problem. ΔM pulse sequence is non-positional (non-weighted), while pulse code modulation (PCM) is a positional (weighted) encoding system. It is believed that the non-positional nature of the ΔM pulse stream can lead to simpler and less expensive digital processing circuits.

In spite of the fast progress of semiconductor technologies, there are still a number of open problems in the area of digital signal processing (DSP). For example, flash converters are bulky and power-hungry. Digital multiplying circuits are bulky and power-hungry as well. This problem becomes more acute when we are dealing with the applications which require a 20-bit resolution or more. The length of the code word can be effectively reduced by using differential PCM. ΔM is one bit DPCM and employs a trade-off between a number of amplitude quantization levels and sampling frequencies. There is a feeling that by increasing the sampling frequency we can reduce the complexity of hardware for arithmetic operations.

In spite of the fact that this approach increases the internal speed of circuitry involved in operations, this compromise is justified for the following reasons:

1. All digital logic families employed in different applications such as control, audio, medical, industrial, etc, operate far below their declared speed. This means that we have at our disposal a "free" frequency band to increase the speed of internal processing.
2. Requirements for complex interfacing filters with ADC are not so strict.
3. With the increased speed of operations, there is an increased possibility of introducing greater "parallelism" for some operations.

As it has been seen in chap. 1, Δ–Σ ADCs are characterized with one bit quantization and a very high sampling rate. To process a Δ–Σ digital stream with ordinary DSP hardware decimation is needed first [1]. The arithmetic operations are then performed with ordinary PCM circuits. Usually these converters are integrated with complex decimation filters on one chip. There are many applications in control, robotics, instrumentation, industrial processes, etc; where direct processing of ΔM pulse stream is preferred.

The advent of low-cost, high-quality Δ–Σ ADCs enabled the development of a new generation of circuits capable of interfacing directly with microprocessors and microcontrollers. There are reports of a 24-bit resolution Δ–ΣM on the market [2]. In the majority of applications, a parallel interface is required between the ADC and the microprocessor. The number of interconnections is directly proportional to the number of parallel bits delivered to the microprocessor. There is need to reduce cost and improve reliability. The answer to this problem is serial interfacing. One of the most important advantages of Δ–ΣM serial output is the possibility of manipulating the serial information in the digital domain, performing linear, non-linear, and mixed-mode operation on a Δ–ΣM pulse stream.

The possibility of direct arithmetic operations on a delta-modulated pulse stream is not fully recognized yet in the literature and practice. The majority of work was done by Kouvaras [5], [4], and Pneumatikakis [5]. In addition to theoretical considerations, the authors of the above references proposed a number of digital circuits for linear arithmetic operation on Δ–Σ modulated pulse streams. Application of ΔM in DSP is relatively new. The majority of publications appeared in the 80's. All of these applications deal mainly with implementation of digital filters with multiplier free coefficients.

## 2.2 EXISTING ARITHMETIC CIRCUITS SOLUTIONS

Addition and multiplication are two basic arithmetic operations involved in the operation of digital filters. In 1972, Lockhart [6] proposed a method of digital filtering using delta modulation. A non-recursive filter is formed by feeding the output of a delta modulator to a binary transferal filter, Fig. 2.1a. The outputs of a serial shift register are weighted according to the impulse response required. For recursive operation, a feedback loop was introduced and a linear delta modulator embedded in a forward path of a recursive system, Fig. 2.1b. The filters discussed by Lockhart do not employ digital arithmetic for coefficient multiplication, and therefore the primary source of noise is introduced by the delta modulator [6].

In reference [7], the use of recirculating shift registers in the implementation of binary transferal filters with quantization coefficients was proposed. The basic arrangement of the proposed solution consists of the use of two recirculating shift registers and a single Exclusive-Or gate. The output of the XOR-gate is led to the analog integrator, Fig. 2.2. The integrator forms the running sum of products [7],

$$\sum_{n=0}^{N} h_n X_{n-r}$$

The binary transferal filter utilizes recirculating shift registers in a serial mode so that only one coefficient multiplier per shift register pair is required. One shift register stores the values of the input signal, while the other shift register stores the values of binary coefficients.

In the last two decades, many papers dealing with the binary arithmetic of digital filters have been published. There are only a limited number of such papers, which deal with applications of delta modulation in digital signal processing. However, a radical approach was made by Peled and Liu [8]. LΔM was used as ADC, and the PCM approach was used for implementation of filter coefficients. Their method for fast filtering was based on read-only memory (ROM), Fig. 2.3.

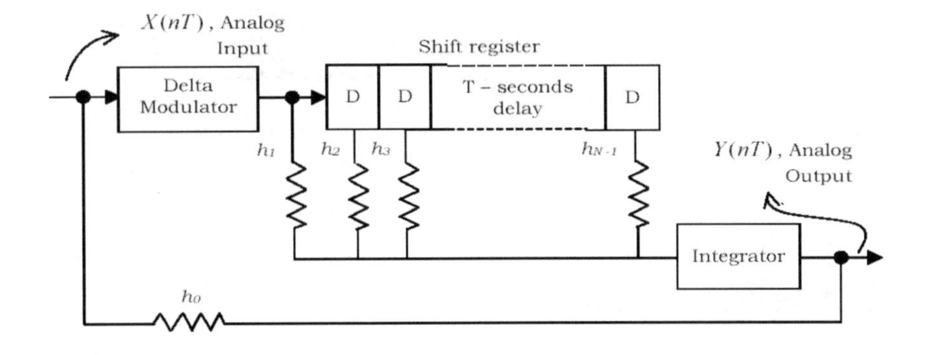


*(a)*

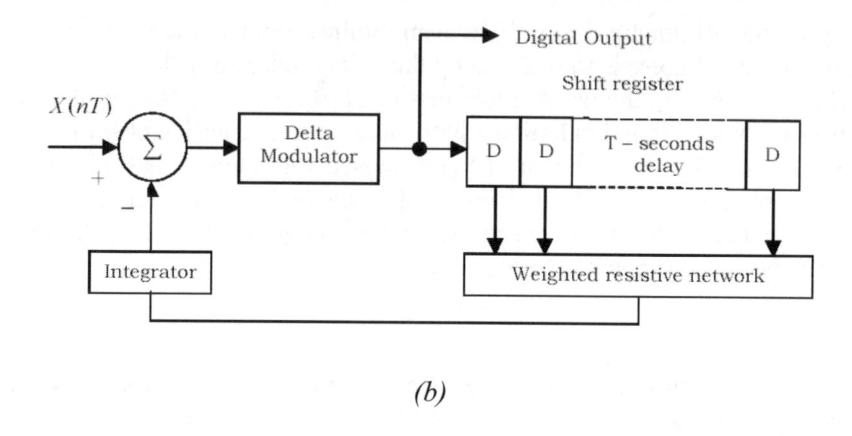


**Fig. 2.1.** Delta modulation filters, (a) non-recursive, (b) recursive [6]

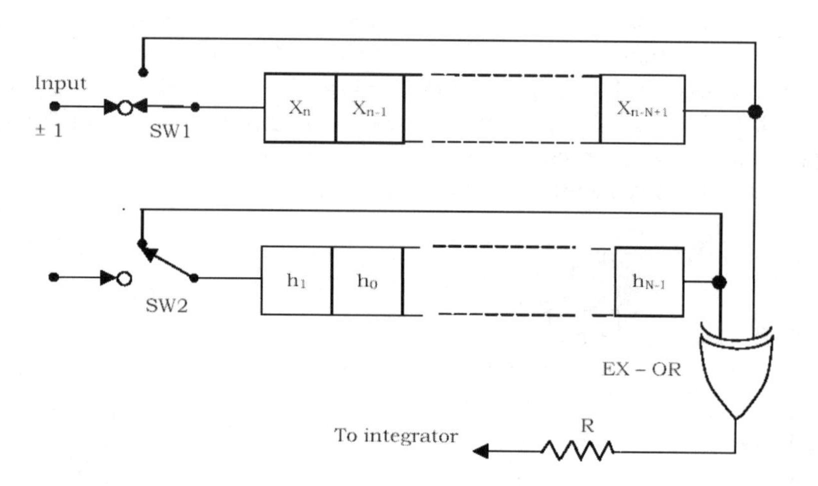


**Fig. 2.2.** Recirculating shift register approach [7]

Peled and Lin avoided the use of adder and multiplier circuits, but they paid the price for this fast structure by large storage. Even in the case of storing a relatively small number of coefficients, the problem is significant. For example, if we wish to implement a filter with 15 coefficients only, the capacity of ROM has to be $2^{15}$ code words. The capacity of ROM can be reduced using the partial sum approach. The idea of partial sum was elaborated more by introducing the "shift and add" principle [9]. The main advantage of this approach was the speed of operation, and reduced complexity and consumption.

A somewhat different approach was proposed by Ashuri [10]. His contribution consists of proposing new processing elements of a binary delta modulated pulse stream. The idea is to convert digital output of ΔM into an "intermediate" binary (IB) stream. In fact, IB represents an averaged original ΔM stream. In addition to a resistive analog adder and subtractor, Ashuri proposed a digital serial adder and subtractor, and an analog multiplier of delta sequences. Unfortunately, the error analysis and performance of proposed elements were not given. Fig. 2.4 shows the circuitry proposed by Ashuri.

Original work in the synthesis of elements for direct processing of the ΔM pulse stream was done by Lockhart [11]. Fig. 2.5 presents a system for addition of two or more ΔM sequences. Using this arrangement, Lockhart demonstrated how divider by K and averager of K delta-modulated inputs can be implemented.

The realization of adaptive delta modulation processors (AΔM) was proposed by Locicero, Schilling and Garodnick [12]. They have shown that signals, which are adaptive delta modulation encoded, can be arithmetically processed directly, without first decoding or converting to pulse code modulation (PCM). They have shown that the sum, difference, and product can be obtained in PCM and AΔM format by operating on the serial ΔM bit stream. For convenience, they used Song audio mode AΔM algorithm [13] in the realization of arithmetic processors. Authors are claiming that proposed designs are general enough to be applied to a large class of digital AΔMs.

A significant amount of research work has been done by Professor Franks and his graduate students. His research proposal to NSF [14] led to significant research results published in the 1980's. The idea for their proposed work was based on previous research done on a digital differential analyzer (DDA) in the period between 1950 and the 1970s [15, 16, 17]. DDA can be considered an optimal structure for the implementation of special purpose signal processing functions [18]. DDA is considered in the literature as a bridge between an analog computer and a general-purpose digital computer [15, 16]. Using the over-sampling technique, a DDA can be employed as a basic building block in many DSP applications.

The DDA circuit element is shown in Fig. 2.6. It consists of two n-bit registers R and Y, transfer device TFR, and a quantizer Q. The role of TFR is to add or subtract the contents of register Y to or from register R, depending on whether $\Delta X(n) = 1$ or $-1$, respectively. If the contents of the R and Y registers are $r(n)$ and $y(n)$, respectively, then the basic idea behind the DDA element is to save the residue of quantization error, $r(n)$. This error is memorized and used in the next computational cycle to compute the new incremental output.

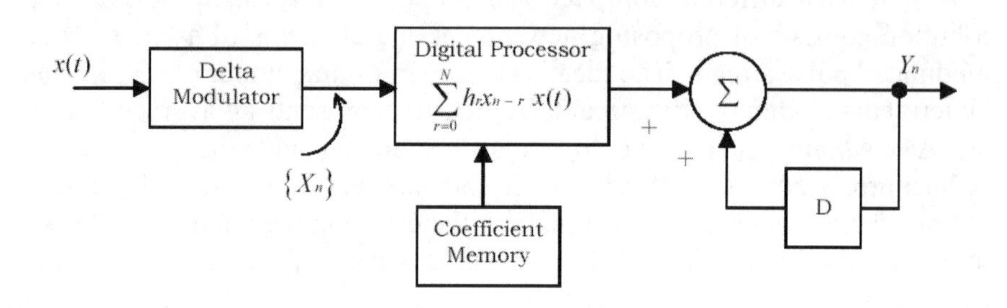


**Fig. 2.3.** ΔM approach of implementation of FIR filter [8]

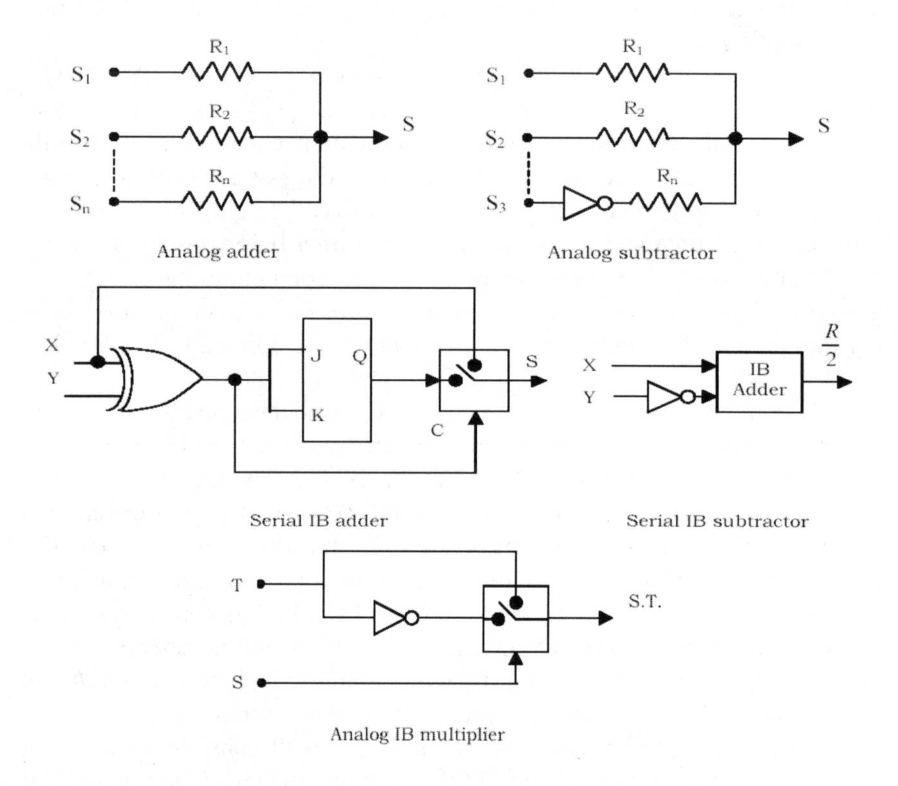


**Fig. 2.4.** Circuits for processing ΔM pulse stream [10]

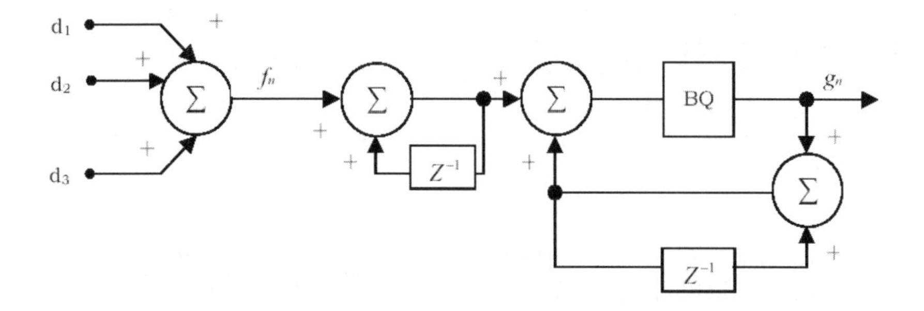


**Fig. 2.5.** *n* – input adder of ΔM signal [11]

To reduce the complexity of ΔM arithmetic circuits, Franks and Hill [14] proposed a delta-sigma modulation adder. The idea is first to decode two ΔM signals, then to add them using an analog adder. The resulting signal is ΔM decoded. Several methods of implementation have been proposed to reduce the complexity of computation circuits using power-of-two filter coefficients [19], [20].

The DDA circuit element is characterized with a low hardwarc complexity and a high level of modularity. Padir [21] analyzed digital incremental recursive filters based upon DDA. Padir studied the idea of multi-bit transfer between DDA modules and analyzed error and limit cycle performance of a first- and second-order, all-pole filter structure. In addition, Padir proposed several bi-quad structures for filter implementation. Much of this information was summarized by Kouvaras [3]. In fact, initial work of Kouvaras has had the biggest influence on the work presented in the chapters that follow. Because of that, the rest of the chapter is dedicated to the summarization of the work of Kouvaras.

### 2.2.1 The Approach of Kouvaras [3]

In his landmark paper [3], Kouvaras introduced a simple and inexpensive method of addition and subtraction of two or more LΔM sequences. The proposed method of arithmetic operations on a LΔM pulse stream is shown in Fig. 2.7. The full treatment and error analysis of the process of addition and multiplication is given in reference [3]. Here we will summarize the main contribution of the paper. Fig. 2.7 shows the process of adding two synchronous linear delta modulated sequences $X_n$ and $Y_n$. They are added in a serial binary full adder. The roles of conventional binary full adder outputs are interchanged. Now the terminal of a carryout C becomes the terminal for the sum $S_n$ of delta sequences, and the terminal of sum S becomes a terminal for the carryout of the delta adder. The resulting signal of the sum is defined as

$$S_n = 2^{-1}\left[x_n + y_n - \left(1 - x_n y_n\right) C_{n-1}\right], \text{ and} \tag{2.1}$$

$$C_n = X_n Y_n C_{n-1} \tag{2.2}$$

where $X_n$, $Y_n$, $C_n$, $C_{n-1}$, and $S_n$ take the values of +1 or -1. From the eqns.s (2.1) and (2.2) Table 2.1 was constructed.

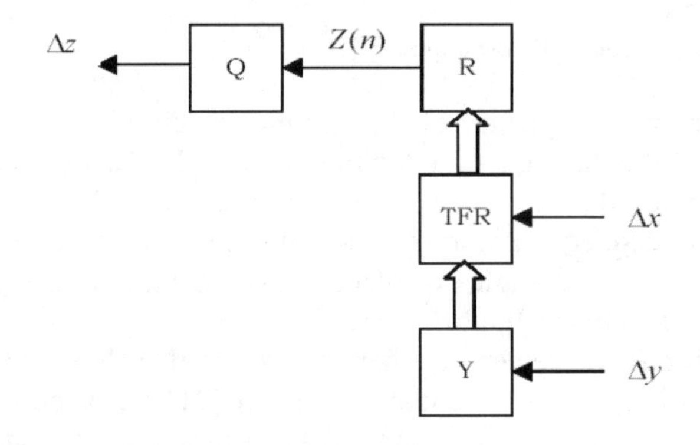


**Fig. 2.6.** The DDA circuit element [14, 15]

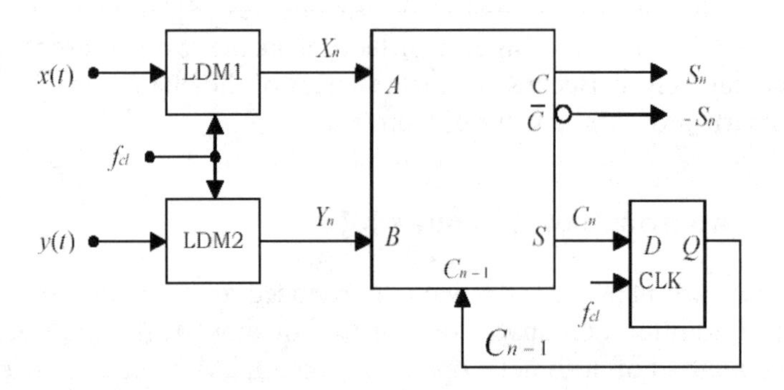


**Fig. 2.7.** Delta full adder

**Table 2.1.**

| $X_n$ | $Y_n$ | $C_{n-1}$ | $S_n$ | $C_n$ |
|---|---|---|---|---|
| -1 | -1 | -1 | -1 | -1 |
| 1 | -1 | -1 | -1 | 1 |
| -1 | 1 | -1 | -1 | 1 |
| 1 | 1 | -1 | 1 | -1 |
| -1 | -1 | 1 | -1 | 1 |
| 1 | -1 | 1 | 1 | -1 |
| -1 | 1 | 1 | 1 | -1 |
| 1 | 1 | 1 | 1 | 1 |

Table 1.1 can be considered a truth table of binary full adder if the value of -1 is substituted with 0 whenever it occurs in it. Replacing -1 with 0, the following logic relations are obtained

$$S_n = X_n Y_n + X_n C_{n-1} + Y_n C_{n-1}, \; C_n = X_n \oplus Y_n \oplus C_{n-1} \tag{2.3}$$

These relations lead Kouvaras to the synthesis of a conventional full adder with the interchange of the roles of the sum and carry outputs, Fig. 2.7. When the binary sequence $S_n$ is demodulated, the one half of the sum of signals $x(t)$ and $y(t)$ is obtained

$$\hat{S}(t) = 2^{-1}\left[x(t) + y(t)\right] - 2^{-1}\left[e_1(t) + e_2(t)\right] + \varphi(t) \tag{2.4}$$

where $2^{-1}[e_1(t) + e_2(t)]$ is the half-sum of the errors of the two LΔM systems and can be considered as the equivalent error of a LΔM system, the input of which is the analog signal $2^{-1}[x_1(t) + x_2(t)]$. The value of $\varphi(t)$, because of the introduction of the full adder, is $|\varphi(t)| \leq \delta$, where $\delta$ is the step size of LΔM. This error can be made smaller with the increase of sampling and the decrease of $\delta$.

With successive application of the operation (2.1), it is possible to find a delta sequence $P_n$, of the product $\alpha x(t)$, where $\alpha$ is a constant with $|\alpha| < 1$. The error due to a multiplication operation is less than $2\delta$ in absolute value.

In addition to the pioneering work mentioned above, Kouvaras proposed several networks for reduction of quantization noise in the direct processing of delta-modulated signals. According to proposed technique [23], the arithmetic network of a digital filter is clocked at a rate higher than that of the delta modulation encoder, i.e.

$$f' = kf, \text{ and } \delta' = \delta/k, \text{ where } k = 1,2,3, \ldots \quad (2.5)$$

If the building elements of the filter arithmetic network are delta adders, then the maximum quantization error is divided by $k$ if the filter has $k$-times the clock rate of the delta encoder. There are two disadvantages concerning filters at the clock rate $f' = kf$:

1. The number of shift register stages is $k$-times that of IIR filters clocked at the rate $f$.
2. The clock rate of output sequence is $k$-times that of input sequence.

To mitigate these disadvantages, Kouvaras proposed a network, which transforms a high rate sequence into an equivalent low rate one. In [24] Kouvaras proposed a multi-input delta signal processing networks. A simple multi-input delta adder was proposed, which gives the same quantization error as two input delta adders.

## 2.3 CONCLUSION

From Kouvaras's work, we concluded that through a direct operation on a delta-modulated pulse stream it is possible to find the half sum of two analog signals. With direct operation on a delta-modulated pulse stream, it was also possible to form a delta-modulated signal of product of an analog signal by a constant smaller than one. The resulting modulated signal includes an error which is dependent on sampling frequency and δ step size. The proposed hardware is simple and modular.

In this chapter, compilation of references relevant to this book is discussed. Significant results of direct processing of a LΔM pulse stream were achieved by N. Kouvaras. Thus, special attention was dedicated to his paper published in 1978 [2]. Chap. 4 of this book presents a generalization of Kouvaras's work for multi-valued delta-modulated signals.

## REFERENCES

1. Norsworthy, S., Schreier, R., and Temes, G., Editors, Delta-sigma Data Converters, IEEE Press, New York, 1997, ISBN: 0-7803-1045.
2. www.analog.com/Sigma-DeltaADCs, AD779x.
3. Kouvaras, N., "Operations on Delta-modulated Signals and their Application in the Realization of Digital Filters", Institute of Electronic and Radio Engineers, Vol. 48, No. 9, pp. 431-438, September 1978.
4. Kouvaras, N., "Modular Network for Direct Complete Addition of Delta-Modulated Signals with Minimum Quantization Error", Int. J. Electronics, 1985, Vol. 59, No. 5, pp. 587-595.
5. Pneumatikakis, A. Deliyannis, T., "Direct Processing of Sigma-Delta Signals", ICECS-1996, pp13-16.
6. Lockhart, G., "Digital Encoding and Filtering Using Delta Modulation", The Radio and Electronic Engineer, Vol. 42m No. 12, December 1972, pp 547-551.
7. Lockhart, G., Babary, S., "Binary Transversal Filters Using Recirculating Shift Registers", The Radio and Electronic Engineer, Vol. 43, No. 3, March 1973, pp. 224-226.
8. Peled, A., Liu, B., "A New Approach to the Realization of Nonrecursive Digital Filters", IEEE Trans. on Audio and Electroacoustics, Vol. AU-21, No. 6, December 1973, pp. 477-484.
9. Peled, A., Lin, B., "A New Hardware Realization of Digital Filters", IEEE Transactions on ASSP, Vol. ASSP-22, No. 6, December 1974, pp. 456-462.
10. Ashouri, M., "A New Approach to the implementation of Signal Processors", Fourth International Symposium on Network Theory, Ljubljana, Slovenia, September 1979, pp. 91-96.
11. Lockhart, G., "Implementation of Delta Modulators for Digital Inputs", IEEE Transactions on ASSP, Vol. ASSP-22, No. 6, December 1974, pp. 453-456.
12. Locicero, J. Schilling, D., Garodnick, J., "Realization of ADM Arithmetic Signal Processors", IEEE Transactions on Communications, Vol. COM-27, No. 8, August 1979, pp 1247-1254.
13. Song, C., Garodnick, J, Schilling, D., "A Variable-step size Robust Delta Modulator," IEEE Transaction on Communication Technology, Vol. COM-19, December 1971, pp 1033-1044.
14. Franks, L., Hill, F., "Simplified Implementation of Digital Signal Processors Based on Delta Modulation", Research proposal, submitted to NSF, March, 1983.
15. Sizer, T., "The Digital Differential Analyzer", Chapman and Hall, Ltd. London, 1968.
16. Schileiko, A., (Translated by D.P. Barret), Digital Differential Analyzers, Pergamon Press, 1964.

17. Maiorov, F., Electronic Digital Integrating Computers – Digital Differential Analyzers, American Elsevier, 1964.
18. Abu-El-Haija, A., Shenoi, K., Peterson, A., "Digital Filter Structures Having Low Errors and Simple Hardware Implementation", IEEE Trans., CAS -25, August 1978, pp. 593-599.
19. Benvenuto, N., Franks, L., Hill, F., "On the Design of FIR Filters with Powers-of-Two Coefficients", IEEE Trans., COM-32, December 1984, pp. 1299-1307.
20. Costas, J., "Computationally Efficient Digital Filters", Proceedings of the IEEE, Vol. 74, Feb. 1986, pp. 371-373.
21. Padir, H., Digital Incremental Recursive Filtering Using Digital Differential Analyzers", Ph.D. Thesis University of Massachusetts, Amherst May 1987.
22. Engel, L. Steenaart, W., "Digital Summation of Delta Modulation Signals", Canadian Communication and Power Conference, Montreal, Canada, 1976, pp. 245-248.
23. Kouvaras, N., "A Technique for a Substantial Reduction of the Quantization Noise in the Direct Processing of Delta-Modulated Signals", Signal Processing, Elsevier Science Publishers, Vol. 8, No. 1, Feb. 1985, pp. 107-119.
24. Kouvaras, N., "Novel Multi-Input Signal Processing Networks with Reduced Quantization Noise", International Journal of Electronics, Vol. 56, No. 3, 1984, pp. 371-378.

# CHAPTER 3 BASIC TERNARY LOGIC CIRCUITS

## 3.1 INTRODUCTION

Our objective is to show that arithmetic operations on multi-valued delta modulated signals are possible as well. Chap. 4 is dedicated to this topic. Thus, we will briefly introduce some basics on multi-valued logic found in reference [1]. The idea of using of multi-valued logic in digital signal processing is not new. It is shown in reference [1] that there exists the "optimal" base number system of R, and with increase of R there is an increase in the amount of information per connection. The question is, what number system gives the most economical implementation of a digital signal processor? A short analysis below shows that the ternary system gives the closest economical solution. Let $d$ represent the maximal number of digits, and R the base of a particular number system. Consequently, the maximum number of different values is $N = R^d$. If we assume that the average cost C, of processing unit for N different values, is directly proportional to the base (radix) R and the maximum number of digits used to represent the number R, then

$$C = K(Rd) = K\left[R\left(\frac{\log N}{\log R}\right)\right] \tag{3.1}$$

where K is a constant.

We are looking for a radix R that gives the minimal cost when N is a constant. Differentiating (3.1) as a function of R and setting the equation to be equal to zero, we get $R = e = 2.7183$. In practice, R must be an integer, thus R = 3. If we assign the cost of the processing unit for a binary system (R = 2) as 100 [1], then Table 3.1 presents the cost of the processing unit for different number systems. It can be seen that the system with $R = 3$ gives an optimal solution. It is fair to say that, with the increase of R, the complexity of a system is increased as well and tolerances become narrower. The natural extension of a binary system is a ternary system, which can be presented as 0, 1, and 2, or -1, 0, and +1. In a ternary system, decimal number D can be represented as follows:

**Table 3.1.**

| Number system R | $C = R.d$ K = 1 | Amount of information per connection |
|---|---|---|
| 2 | 100.0 | 1.000 |
| 3 | 95.0 | 1.585 |
| 4 | 100.0 | 2.000 |
| 5 | 107.9 | 2.322 |
| 10 | 150.5 | 3.322 |

$$D = \{T_n 3^n + T_{n-1} 3^{n-1} + \ldots + T_1 3^1 + T_0 3^0\} \tag{3.2}$$

where

T = ternary digit from the set: {0, 1, 2} or {-1, 0, +1}

n = the weight of ternary digits

$T_0$ = the least significant ternary digit

$T_n$ = the most significant ternary digit.

In our consideration, a symmetrical ternary system {-1, 0, 1} is adopted. There are certain advantages of using a symmetrical ternary system. First, any number can change its positive value into negative by substituting +1 with -1, and vice versa. The sign of a most significant digit tells whether the number is positive or negative. Zero values stay unchanged. Operation of addition and subtraction is implemented with the same hardware only with the change of the sign of the number, which is added or subtracted. "Carry out" circuits are unnecessary because the numbers are rounded up to the most significant digit with the transfer of the remaining digit [2], [3].

The price of an integrated circuit is directly proportional to the number of connections and interconnections in a system or a subsystem. Research results of Vranesic [2] show that the number of interconnecting wires in a ternary parallel multiplier is 2/3 less than in an equivalent binary configuration. The same author proves that the number of circuits is reduced by nearly 20%.

Since signal processing of multi-valued delta-modulated pulse streams is less known in practice than signal processing of a binary delta modulated pulse stream, a brief description of some ternary logic elements is given first. For the purposes of this book, some basic knowledge of ternary full adder and ternary D-flip-flop is needed.

## 3.2 MULTI-VALUED ALGEBRA AND FUNCTIONAL COMPLETENESS

For the multi-valued logic system presented in Fig. 3.1, the number of different input combinations is $R^n$ and the number of different functions at the output is $R^{(R)}$.

For a three-level system, there are $3^n$ input combinations and $3^3$ ternary functions for three input variables. A trivial case $f(x)$ = *constant* and all degenerative functions, with less than $n$ inputs, are included. For only one ternary input $X_i$, there exist 27 possible ternary functions $f(X_i)$. Table 3.2 shows the number of possible functions for both the binary and the ternary case.

For the ternary system to be fully functional and complete, all $3^n$ functions must be synthesized. There are several tests to prove whether the given algebra is functionally complete or not. In the case of binary logic, it is enough to have OR, AND, and inverter gates to implement any boolean logic function. Similarly, in the case of multi-valued logic, it is necessary to have algebra and multi-valued circuits to implement all functions except a constant. In this case, it is said that algebra is functionally complete. There are several test methods to examine whether an algebra, for $R > 2$, is functionally complete or not [1], for example, post-algebra, modulo-algebra, single operator algebra, hardware oriented algebra, etc.

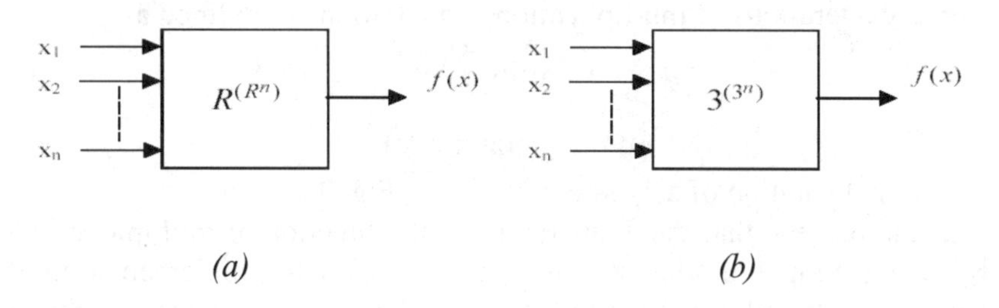


**Fig. 3.1.** Multi-valued system with one output, (a) system with R levels, (b) system with three levels

**Table 3.2.**

| | Binary | | Ternary |
|---|---|---|---|
| $X_i$ | $f(X_i)$ | $X_i$ | $f(X_i)$ |
| 0 | 0 0 1 1 | -1 | -1-1-1-1-1-1-1-1-1 0 0 0 0 0 0 0 0 0 1 1 1 1 1 1 1 1 1 |
| 1 | 0 1 0 1 | 0 | -1-1-1 0 0 0 1 1 1-1-1-1 0 0 0 1 1 1-1-1-1 0 0 0 1 1 1 |
| | | 1 | -1 0 1-1 0 1-1 0 1-1 0 1-1 0 1-1 0 1-1 0 1-1 0 1-1 0 1 |

For the purposes of this book, algebra from Lee and Chen is used [1], [4]. Their proposal consists of basic ternary operator T with fewer inputs. This operator is defined as

$$T(p,q,r) = \begin{cases} p, & if\ s = 1 \\ q, & if\ s = 0 \\ r, & if\ s = -1 \end{cases} \tag{3.3}$$

where $p$, $q$ and $r$ are 1, 0, and -1 respectively, and s is select input. According to [4] ternary switching function $T(p,q,r;s)$ is defined as

$$\mathrm{T}(\mathrm{p},\mathrm{q},\mathrm{r};\mathrm{s}) = \mathrm{pI}_1(s) + qI_0(s) + rI_{-1}(s) \tag{3.4}$$

$I_k(s)$ is defined as

$$I_k(s) = \begin{cases} 1, & s = k \\ -1, & s \neq k \end{cases} \tag{3.5}$$

Ternary operations of multiplication and addition are defined as

$$x \cdot y = \min(x, y) \tag{3.6}$$

$$x + y = \max(x, y)$$

Functional solution of a T-gate is shown in Fig. 3.2.

It can be seen that the T operator has the function of multiplexer where S is control input or address. Discrete semiconductor implementation of T-gate can be found in reference [4] or in [5]. More details and insight about multi-valued logic trend and development can be found is reference [6].

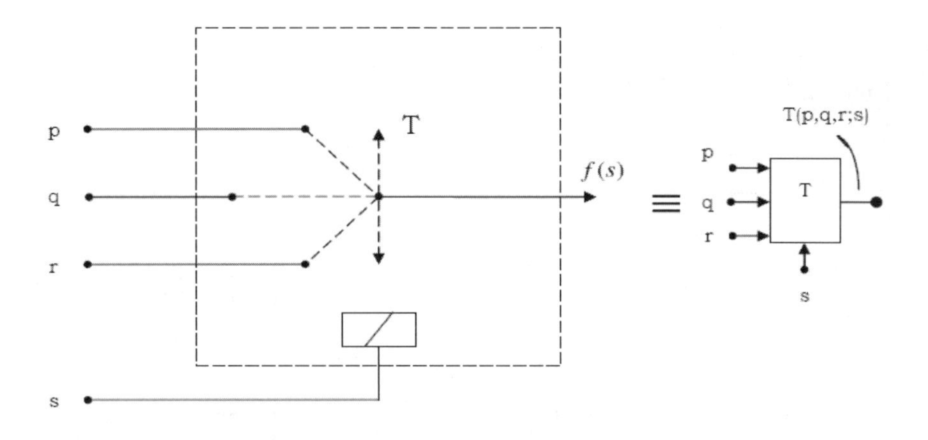


**Fig. 3.2.** Lee's and Chen's functionally complete operator T

## 3.3 IMPLEMENTATION OF TERNARY FULL ADDER

It is possible to implement 27 unary functions with one argument using only one T-gate. Combining more T-gates is possible to implement ternary full adder. This adder will be needed for implementation of ternary delta adder in chap. 4. Table 3.3 shows the truth table of ternary full adder.

The sum and carry-out are expressed as

$$S(a,b,c) = T(T(0,-1,1;a),T(-1,1,0;a),a;a') \tag{3.7}$$

$$a' = T(b,T(0,-1,1;b),T(-1,1,0;b);c) \tag{3.8}$$

$$C(a,b,c) = T\begin{pmatrix} T(T(1,1,0;a),T(1,0,0;a),0;b), \\ T(T(1,0,0;a),0,T(0,0,-1;a);b), \\ T(0,T(0,0,-1;a),T(0,-1,-1;a);b);c \end{pmatrix}. \tag{3.9}$$

Using eqns. (3.7), (3.8), and (3.9) ternary full adder is synthesized, Fig. 3.3. Then serial ternary full adder is implemented when carry out is delayed using ternary D flip-flop [4].

**Table 3.3.**

| | S(a,b,c) |
|---|---|
| a \ b,c | 1 0 -1 |
| 1 1 | 0 -1 1 |
| 1 0 | -1 1 0 |
| 1 -1 | 1 0 -1 |
| 0 1 | -1 1 0 |
| 0 0 | 1 0 -1 |
| 0 -1 | 0 -1 1 |
| -1 1 | 1 0 -1 |
| -1 0 | 0 -1 1 |
| -1 -1 | -1 1 0 |

| | C(a,b,c) |
|---|---|
| a \ b,c | 1 0 -1 |
| 1 1 | 1 1 0 |
| 1 0 | 1 0 0 |
| 1 -1 | 0 0 0 |
| 0 1 | 1 0 0 |
| 0 0 | 0 0 0 |
| 0 -1 | 0 0 -1 |
| -1 1 | 0 0 0 |
| -1 0 | 0 0 -1 |
| -1 -1 | 0 -1 -1 |

Fig. 3.4 presents an example of the operation of a ternary serial full adder. *(a,b)/S* are inputs and output respectively. A, B and C are internal states of the adder, i.e., the states of "carry" 1, 0, -1. Fig. 3.4a shows a transition diagram of a serial ternary adder as a function of periodic input sequence S. Fig. 3.4b shows corresponding waveform signals.

## 3.4 MEMORY ELEMENT BASED ON T-GATE

For synthesis of serial ternary full adder ternary logic, a delay element is needed as well. Fig. 3.5 presents a block diagram of ternary D flip-flop with corresponding waveforms [4].

As can be seen from Fig. 3.5, for proper operation of this FF two clock pulses are needed. They can be implemented by differentiating the basic clock pulse CP. Fig. 3.5 b) shows the case of shifting operation when ternary signal P is applied. In conclusion, we can state that the T-gate is successfully used for synthesis of sequential circuits as well. T-gate is considered to be a universal logic module [8]. The objective of this chapter was to introduce the basic principles of a ternary full adder, which will be needed for the synthesis of the ternary delta adder introduced in chap. 4. We advise readers who are interested in this topic to visit the conference proceedings on multi-valued logic held in the past thirty years.

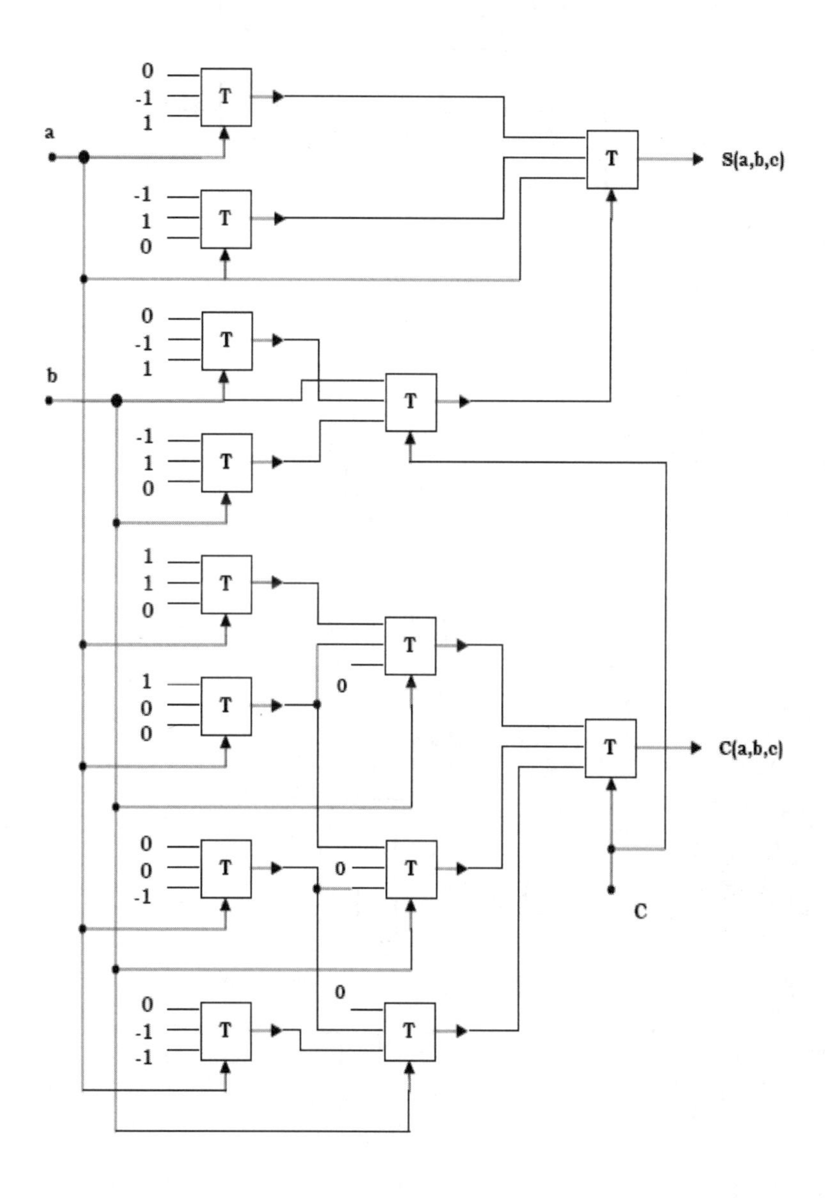


**Fig. 3.3.** Block diagram of a ternary full adder

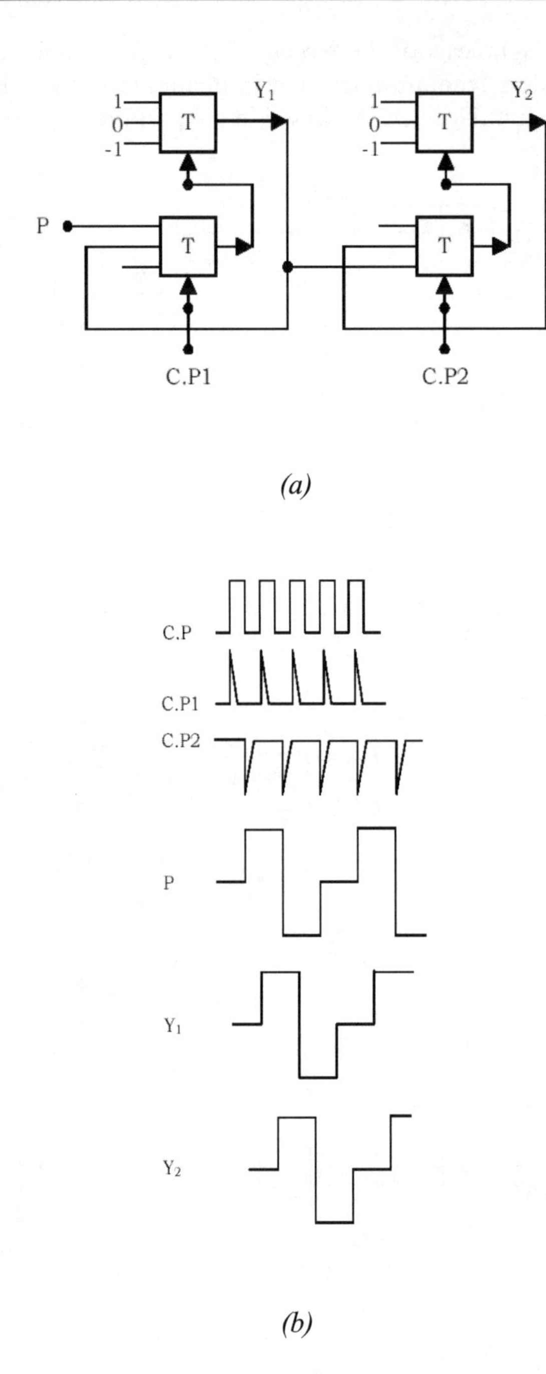


**Fig. 3.4.** Ternary D flip-flop with corresponding waveforms

## 3.5 CONCLUSION

The objective of this chapter was to introduce the basic concept of multi-valued logic. We briefly introduced ternary full adder, whose operations are based on a ternary (T) gate. This introduction will prove helpful to understand the next chapter.

## REFERENCES

1. Hurst, L., The Logic Processing of Digital Signals, Crane Russak and Company Inc., New York, N.Y., 1978.
2. Vranesic, Z., Smith, K., Engineering Aspects of Multi Valued Logic Systems, Computer, September 1974, pp. 34-41.
3. Avizienis, A. "Signed-Digit Number Representation for Fast Parallel Arithmetic", IRE Tr. On electronic Computers, September 1961, pp. 389-400.
4. Higuchi, T., Kameyama, M., Ternary Logic Systems based on T-Gate, International Symposium on Multi-valued Logic, 1975, pp. 290-304.
5. Porat, D., "Three-Valued Digital System", Proceedings of IEEE, Vol. 166, pp. 947-954, June 1969.
6. IEEE Transaction on Computers, Vol. C-30, No. 9, September 1981.
7. Yan, S., Tang, C., "Universal Logic Modules and Their Applications", IEEE Transaction on Computers, Vol. C-19, pp. 141-149, February 1970.
8. Higuchi T., Kameyama, M., Synthesis of Multiple-Valued Universal Logic Modules", Intern. Symposium on Multi Valued Logic, pp. 121-130, 1975.

# CHAPTER 4 MULTIVALUED ARITHMETIC OPERATIONS

## 4.1 INTRODUCTION

As stated in chap. 3, binary logic is not always optimal and significant simplification in hardware implementation can often be achieved by going to a higher multi-valued logic [1]. Specifically, in VLSI technology, the major problems are associated with the complexity of the connections between components on a chip and the interconnections between chips. The cost of components is a relatively small part of the total price of the system. More and more, the complexity of interconnections between subsystems dictates the overall cost of the system.

Connectors are expensive. They introduce the usual noise and reliability problems, and require expensive testing. By increasing the information rate per wire, one can reduce the number of wires in a digital filter without reducing the amount of information transmitted. The application of m-valued logic and the use of ΔM conversion (LΔM or Δ-ΣM) in digital signal processing is one way of increasing the information rate per wire. The objective of this chapter is to show that arithmetic operations on ΔM pulse stream are possible.

## 4.2 ADDITION OF TWO OR MORE TΔM SEQUENCES

To define addition of two or more TΔM signals, we will follow the approach similar to Kouvaras [2]. Kouvaras has shown that binary full adder can be used for implementation of addition and multiplication with a constant less than one. Similarly, the ternary full adder presented in chap. 3 can be employed for addition of TΔM signals. The operations of binary and ternary full adders are different, thus generalization is not so trivial.

Let us consider implementation of a ternary delta adder (TΔA) shown in Fig. 4.1. It consists of a ternary full adder and a D flip-flop, Fig. 3.3 and 3.4. These two elements are connected according to reference [3]. The only difference is that outputs S and C have interchanged roles. Fig. 4.1b is a symbolic presentation of TΔA.

Let $\{X_n\}$ and $\{Y_n\}$ be output sequences of identical TΔMs which have analog inputs $x(t)$ and $y(t)$, respectively, and which are controlled by the same clock generator. The ternary sequence is

$$\{S_n\} = \{X_n, Y_n\} \tag{4.1}$$

which will from now on be termed as ternary delta sequence (TΔSQ) of the sum of $\{X_n\}$ and $\{Y_n\}$. It was easy to define $\{S_n\}$ for the binary case [2]. Unfortunately for the ternary case, we have $3^3 = 27$ different combinations, which describe the operation of TΔA. In the case of quaternary delta modulation, the problem becomes even more difficult. As will be shown in this chapter, using quaternary redundant symmetric system, the sign-digit numbers are represented by values from the set {-3, -2, -1, 0, 1, 2, 3}. Expression for the sum has 147 different terms. A different way is needed than that proposed by Kouvaras for defining expression for sum and carry out of multi-valued delta adder. In reference [4], the Lagrange polynomial was used to evaluate values for sum and carry out of multivalued delta adder.

### 4.2.1 Addition of Two Ternary ΔM Sequences

Fig. 4.2 presents the case of adding two synchronous ternary delta modulated sequences and Table 4.1 presents the truth table of ternary full adder.

Our goal is to show that ternary full adder can be used for addition of ternary delta modulated sequences as well. It is possible to see from Table 4.1 that direct application of the Lagrange interpolation formula is complicated. Application of this formula, for calculation of $S_n$ as function of $X_n, Y_n$, and $C_{n-1}$, will give eight-term polynomial of a 26$^{th}$ order. Expression for $C_n$ will consist of 18 terms. This kind of expression can complicate analysis. Having in mind that $S_n$ represents the sum of three numbers ($X_n$, $Y_n$, and $C_{n-1}$), thus $S_n$ must be a function of

$$Z_n = X_n + Y_n + C_{n-1} \tag{4.2}$$

Using this formula, table 4.1 becomes as in table 4.2. Now we can apply the Lagrange interpolation formula to get expressions for $S_n$ and $C_n$ as a function of $Z_n$. The general form of this formula [5] can be written as

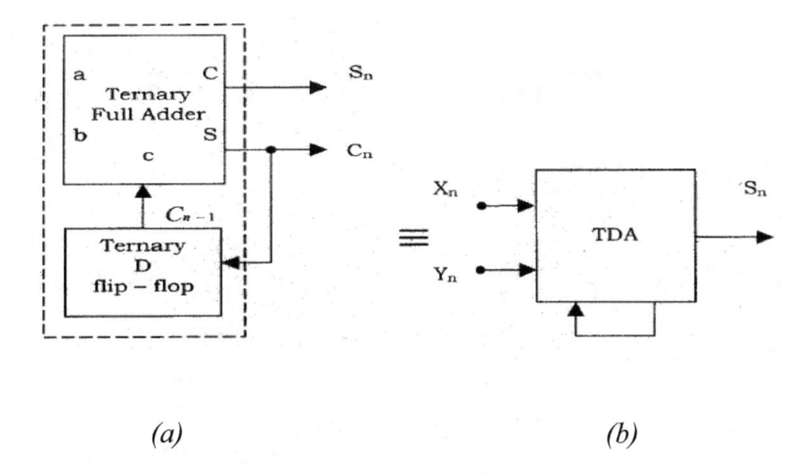


**Fig. 4.1.** (a) Block diagram of TΔA, (b) symbolic representation

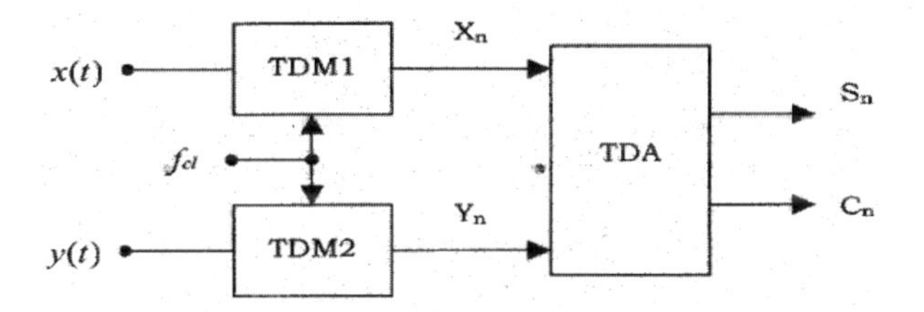


**Fig. 4.2.** System for addition for two ternary ΔM sequences

$$S_n(z) = \sum_{k=1}^{N} g_k L_k^{(n)}(z) \tag{4.3}$$

where

$$L_k^n(z) = \prod_{i=1}^{N} \frac{(z - z_i)}{(z_k - z_i)} \tag{4.4}$$

and

$$g_k = S_n(z_k) \tag{4.5}$$

We can see from table 4.2 that $g_k$ has four values different than zero. Thus,

$$S_n(z_n) = (1)\frac{(z_n-2)(z_n-1)(z_n-0)(z_n+1)(z_n+2)(z_n+3)}{(3-2)(3-1)(3-0)(3+1)(3+2)(3+3)} + \tag{4.6}$$
$$+(1)\frac{(z_n-3)(z_n-1)(z_n-0)(z_n+1)(z_n+2)(z_n+3)}{(2-3)(2-1)(2-0)(2+1)(2+2)(3+3)} +$$
$$+(-1)\frac{(z_n-3)(z_n-2)(z_n-1)(z_n-0)(z_n+1)(z_n+3)}{(-2-3)(-2-2)(-2-1)(-2-0)(-2+1)(-2+3)} +$$
$$+(-1)\frac{(z_n-3)(z_n-2)(z_n-1)(z_n-0)(z_n+1)(z_n+2)}{(-3-3)(-3-2)(-3-1)(-3-0)(-3+1)(-3+2)}$$

After relatively simple manipulation, we get

$$S_n = \frac{-3z_n^5 + 35z_n^3 - 32z_n}{120} \tag{4.7}$$

Applying the same procedure for $C_n$ we have

$$C_n = \frac{3z_n^5 - 35z_n^3 + 72z_n}{40}. \tag{4.8}$$

It is easy to see, using (4.7) and (4.8) that

$$S_n = \frac{z_n - C_n}{3}. \tag{4.9}$$

Finally we can write the expression for $S_n$ as a function of $X_n$, $Y_n$, and $C_{n-1}$

$$S_n = \frac{1}{3}\left(X_n + Y_n + C_{n-1} - C_n\right). \tag{4.10}$$

According to figs 1.16 and 4.2, the relation between modulated and demodulated waveforms can be written as

$$x(t) = \hat{x}(t) + \varepsilon_x(t),\ y(t) = \hat{y}(t) + \varepsilon_y(t),\ \text{for } nT \le t < (n+1)T. \tag{4.11}$$

Corresponding signals for ternary delta sequence are $s(t)$, $\hat{s}(t)$ and $\varepsilon_s(t)$

$$s(t) = \hat{s}(t) + \varepsilon_s(t) \tag{4.12}$$

and

**Table 4.1.**

| IN | | | OUT | |
|---|---|---|---|---|
| $C_{n-1}$ | $X_n$ | $Y_n$ | $C_n$ | $S_n$ |
| -1 | -1 | -1 | 0 | -1 |
| -1 | -1 | 0 | +1 | -1 |
| -1 | -1 | +1 | -1 | 0 |
| -1 | 0 | -1 | +1 | -1 |
| -1 | 0 | 0 | -1 | 0 |
| -1 | 0 | +1 | 0 | 0 |
| -1 | +1 | -1 | -1 | 0 |
| -1 | +1 | 0 | 0 | 0 |
| -1 | +1 | +1 | +1 | 0 |
| 0 | -1 | -1 | +1 | -1 |
| 0 | -1 | 0 | -1 | 0 |
| 0 | -1 | +1 | 0 | 0 |
| 0 | 0 | -1 | -1 | 0 |
| 0 | 0 | 0 | 0 | 0 |
| 0 | 0 | +1 | +1 | 0 |
| 0 | +1 | -1 | 0 | 0 |
| 0 | +1 | 0 | +1 | 0 |
| 0 | +1 | +1 | -1 | +1 |
| +1 | -1 | -1 | -1 | 0 |
| +1 | -1 | 0 | 0 | 0 |
| +1 | -1 | +1 | +1 | 0 |
| +1 | 0 | -1 | 0 | 0 |
| +1 | 0 | 0 | +1 | 0 |
| +1 | 0 | +1 | -1 | +1 |
| +1 | +1 | -1 | +1 | 0 |
| +1 | +1 | 0 | -1 | +1 |
| +1 | +1 | +1 | 0 | +1 |

**Table 4.2.**

| $Z_n$ | 3 | 2 | 1 | 0 | -1 | -2 | -3 |
|---|---|---|---|---|---|---|---|
| $S_n$ | 1 | 1 | 0 | 0 | 0 | -1 | -1 |
| $C_n$ | 0 | -1 | 1 | 0 | -1 | 1 | 0 |

$$\hat{s}(nT) = \delta \sum_{k=-\infty}^{n-1} S_k \text{ , for } nT \le t < (n+1)T \tag{4.13}$$

Plugging (4.10) into (4.13), we have

$$\delta \sum_{l=k}^{n-1} S_l = \frac{1}{3}\delta \sum_{l=k}^{n-1} X_i + \frac{1}{3}\delta \sum_{l=k}^{n-1} Y_i + \frac{1}{3}\delta \sum_{l=k}^{n-1} (C_{l-1} - C_l) \tag{4.14}$$

The error function $\varphi(nT)$ is defined as

$$\frac{1}{3}\delta \sum_{l=k}^{n-1} (C_{l-1} - C_l) = \frac{\delta}{3}(C_{k-1} - C_{n-1}) = \varphi_k(nT) \tag{4.15}$$

For the case when $k \to \infty$, $\varphi_k(nT) = \varphi_n(nT)$. We can write,

$$\hat{s}(nT) = \frac{1}{3}\left[\hat{x}(nT) + \hat{y}(nT)\right] + \varphi(nT) \tag{4.16}$$

or

$$\hat{s}(t) = \frac{1}{3}\left[\hat{x}(t) + \hat{y}(t)\right] + \varphi(t) \text{, for } nT \le t < (n+1)T \tag{4.17}$$

We can conclude from (4.17) that the demodulated waveform is equal to one third of the sum of demodulated signals $\hat{x}(t)$ and $\hat{y}(t)$ plus some error $\varphi(t)$. Having in mind that $C_i$ can have values from the set {*-1, 0, +1*}, the eqn. (4.15) can be written as

$$|\varphi(t)| \le \frac{2}{3}\delta \tag{4.18}$$

The error of sum, using eqns. (4.11), (4.12) and (4.17) is

$$\varepsilon_s(t) = \varphi(t) - \frac{1}{3}\left[\varepsilon_x(t) + \varepsilon_y(t)\right] \tag{4.19}$$

$$| \, \varepsilon_s(t) \, | = | \, \varphi(t) \, | + \frac{1}{3} | \, \varepsilon_x(t) \, | + \frac{1}{3} | \, \varepsilon_y(t) \, | \tag{4.20}$$

because $\varepsilon_x(t)$ and $\varepsilon_y(t)$ are proportional to the $\delta$ step size, thus the total error of summation is

$$\varepsilon_s(t) < k_s \delta \tag{4.21}$$

where $k_s$ is a constant. We can see that this error can be minimized with proper selection of δ and a corresponding increase of sampling frequency such that product $\delta T$ remains constant. From the equations above, we can conclude that the ternary delta adder can be used for addition of two or more ternary delta modulated sequences [6].

### 4.2.2 Addition of Several Ternary ΔM Sequences

If $x_1(t), x_2(t), ..., x_l(t)$ represent $l$ analog signals, then the TΔSQ's of these individual signals determine TΔSQ of the sum

$$s(t) = \frac{1}{3^{r+1}} \left[ \sum_{i=1}^{l} x_i(t) \right] \tag{4.22}$$

where $r$ is the positive integer satisfying $2^r < l < 2^{r+1}$, for $r = 1,2,...n$. The TΔSQ of the sum $s(t)$ can in fact be determined with absolute error less than $(2/3)(r+1)\delta T$. The error growth as a function of $r$ is the result of the successive summing. For example, if $l = 2^3$, then $r = 2$. The successive groupings shown in Fig. 4.3 demonstrate that $s(t)$ can be obtained in these steps with an absolute error less than $(2/3)\delta T$ per step, i.e. with an absolute total error less than $(2/3)3\delta T = 2\delta T$ [6]. If $l$ is not an exact power of 2, the gaps must be filled with idling sequences $x_i(t) = I_n$ as defined in [2].

Fig. 4.4a shows the block diagram of the simulation. Fig. 4.4b shows the reconstruction signal $\hat{x}(t)$ and the reconstruction signal of a sum $\hat{s}(t)$.

TΔM and TΔ demodulator are connected back-to-back. This is the unfiltered sum of inputs $x(t) + y(t)$, where $y(t) = 0$, and $x(t)$ is sinusoid of frequency 10 Hz. Signal $\hat{s}(t)$ presents the unfiltered sum after the demodulation of the ternary sequence $S_n$, and we can see that its amplitude is one third of the sum $x(t) + y(t)$. Delta step size is chosen to be $\delta = 0.001$, and the number of samples per period of signal is 64.

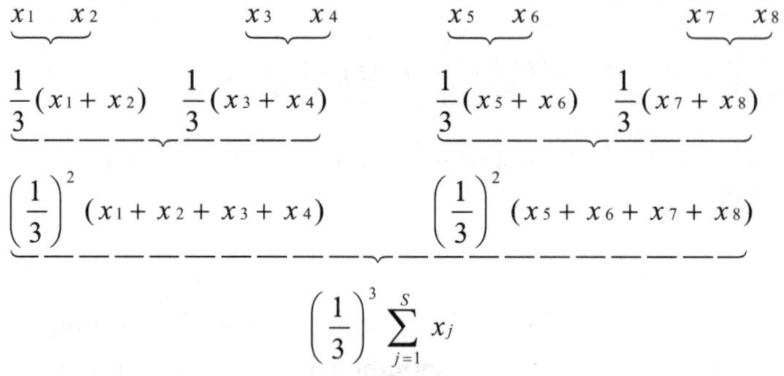

x1 x2 x3 x4 x5 x6 x7 x8
1/3 (x1 + x2) 1/3 (x3 + x4) 1/3 (x5 + x6) 1/3 (x7 + x8)
(1/3)^2 (x1 + x2 + x3 + x4) (1/3)^2 (x5 + x6 + x7 + x8)
(1/3)^3 Σ_{j=1}^{S} xj

**Fig. 4.3.** Successive grouping

x(t)
y(t)
TDM1
TDM2
Xn
Yn
x̂(t)
ŷ(t)
TDM -1
TDA
S
C
Cn
Sn
ŝ(t)
(a)
x^(t)
s^(t)
x^(t), s^(t)
t (sec)
0.1
0.05
0
-0.05
-0.1
0.8
1
1.2
1.4
1.6
1.8
2
2.2
2.4
(b)

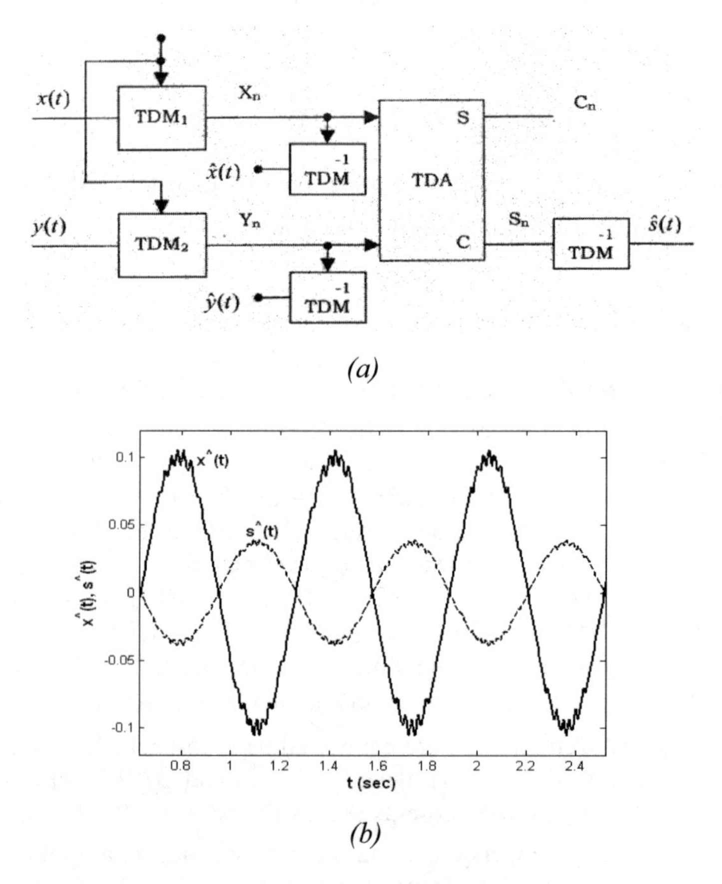

**Fig. 4.4.** (a) block diagram of a simulation model, (b) reconstructed signal $\hat{x}(t)$ and reconstructed signal of the sum $\hat{s}(t)$

In summary, we have shown that a ternary delta adder can be implemented using a conventional ternary full adder when the sum and carry out terminals interchange their roles. We have shown that during the addition operation, multiplication by a constant of 1/3 is introduced. This property of a ternary adder can be efficiently used for multiplications of TΔM signals by constants less than 0.5. In the text that follows, operation of multiplication will be described.

### 4.2.3 Multiplication of TΔM Signal With a Constant

Let α denote a constant of multiplication, $\{X_n\}$ the input ternary delta sequence, and $\{Q_n\}$ the output ternary sequence of delta multiplier. We shall derive an expression for $\{Q_n\}$ a ternary delta sequence corresponding to the product $\alpha \cdot x(t)$, where $x(t)$ is decoded signal of the ternary delta sequence $\{X_n\}$. The constant α is assumed different from zero and is explicitly given in the form

$$\alpha = \sum_{j=1}^{q} \alpha_j 3^{-j} \,, \alpha_j = \{-1, 0, +1\}. \tag{4.23}$$

If $\alpha_j = 1$ for all $j = 1, 2, \ldots, q$, the restriction

$$\alpha \le \frac{1}{3} + \frac{1}{3^2} + \ldots + \frac{1}{3^q} = \frac{1}{3}\left[1 - \left(\frac{1}{3}\right)^q\right] \tag{4.24}$$

holds for all $q$, so we have $\alpha \le (1/2)$ when $q \to \infty$. Let us define the ternary sequence $\{Q_n\}$ corresponding to the product $\alpha x(t)$ as in [6] via

$$\{Q_n\} = \left\{B_n^{(1)}, \ldots, \left\{B_n^{(q-1)}\right\}, \left\{B^q, I_n\right\} \ldots\right\} \tag{4.25}$$

where

$$\left\{B_n^{(j)}\right\} = \{X_n\}, \text{ if } \alpha_j = +1 \tag{4.26}$$

$$\left\{B_n^{(j)}\right\} = \{-X_n\}, \text{ if } \alpha_j = -1$$

$$\left\{B_n^{(j)}\right\} = \{I_n\}, \text{ if } \alpha_j = 0$$

$I_n$ represents the idle sequence defined as $I_n = 0$, for n = …-2,-1,0,1,2,…
If we use the expression for $\{S_n\}$ as

$$S_n = \frac{1}{3}\left[X_n + Y_n - (C_n - C_{n+1})\right] \tag{4.27}$$

then considering eqn. (4.10), the individual sums of $\{S_n\}$ can be written as

$$S_n^{(1)} = \frac{1}{3}\left[B_n^{(q)} + I_n - \left(C_n^{(1)} - C_{n+1}^{(1)}\right)\right] \tag{4.28}$$

$$S_n^{(2)} = \frac{1}{3}\left[\frac{1}{3}\left(B_n^{(q)} + I_n - \left(C_n^{(1)} - C_{n+1}^{(1)}\right)\right) + B_n^{(q-1)} - \left(C_n^{(2)} - C_{n+1}^{(2)}\right)\right]$$

$$\vdots$$

$$S_n^{(q)} = \left(B_n^{(1)}, S_n^{(q-1)}\right) = Q_n$$

From eqn. (4.28), we see that $S_n^{(q)}$ represents $Q_n$ as well. So we obtain,

$$Q_n = \sum_{j=1}^{q} 3^{-j} B_n^{-j} + 3^{-q} I_n - (k_n - k_{n-1}), \tag{4.29}$$

for $n = \ldots, -2, -1, 0, +1, +2, \ldots$, where

$$k_n = \sum_{j=1}^{q} \left(C_n^{(j)}\right)\left(3^{j-1-q}\right) \tag{4.30}$$

$C_n^{(j)}$ corresponds to the carry output in the process of forming the partial sum $S_n^{(j)}$ $(j = 1, 2, \ldots, q)$, the expressions (4.25), (4.27), (4.28), and (4.29) are important in the implementation of the ternary delta multiplier.

### 4.2.4 Synthesis of Ternary Delta Multiplier

Let $\{X_n\}$ represent a ternary delta sequence corresponding to the analog signal $x(t)$, which is multiplied by the constant $\alpha$ given by,

$$\alpha = \sum_{j=1}^{q} \alpha_j 3^{-j} \tag{4.31}$$

where $\alpha_j \in \{-1, 0, +1\}$, $(j = 1, 2, \ldots, q-1)$, $\alpha_q \in \{-1, +1\}$. Then $\{Q_n\}$, as defined in (4.29), represents a ternary sequence corresponding to the analog signal $\alpha . x(t)$ with an error $\alpha(t) = F_{DM}^{-1}(k_n - k_{n-1})$, where $k_n$ is defined by (4.30) and

$$F_{DM}^{-1} = \alpha \sum_{k=-\infty}^{\infty} X_k \tag{4.32}$$

is the operator which transforms (demodulates) the ternary delta sequence into an analog signal. We claim that the restriction $\sigma_{mx}(t) \leq [1 - (1/3)^q]\delta$ holds.

**Proof:** The expression for $B_n^{(j)}$ can be written as follows

$$B_n^{(j)} = \alpha_j X_n + (1-\alpha_j)(1+\alpha_j)I_n, \mathrm{j} = 1,2, \ldots, \mathrm{q} \tag{4.33}$$

Inserting (4.33) into (4.29) we obtain,

$$Q_n = \sum_{j=1}^{q} 3^{-j}\left[\alpha_j X_n + (1-\alpha_j)(1+\alpha_j)I_n\right] + 3^{-q} I_n - (k_n - k_{n-1}) \tag{4.34}$$

or

$$Q_n = X_n \sum_{j=1}^{q} 3^{-j}\alpha_j + I_n \sum_{j=1}^{q} 3^{-j}(1-\alpha_j^2) + 3^{-q} I_n - (k_n - k_{n-1}) \tag{4.35}$$

because $I_n$ is defined to be zero. It is evident that $I_n$ can be omitted. After demodulating sequence $\{Q_n\}$, we obtain

$$\hat{Q}(t) = \delta \sum_{k=-\infty}^{n} Q_k = \alpha \cdot \hat{x}(t) - F_{DM}^{-1}(k_n - k_{n-1}), \tag{4.36}$$

for $nT \leq t \leq (n+1)T$. The idle sequence, multiplied by a constant factor, becomes equal to zero after the $F_{DM}^{-1}$ operation. For the estimation of the maximum error magnitude $\sigma(t) = F_{DM}^{-1}(k_n - k_{n-1})$, it is necessary to take into account the error introduced by the ternary delta adder [6]. Consider the partial sum given in eqn. (4.28). The error introduced in the first partial sum is $(2/3)\delta$. This error propagates in the partial sum $S_n^{(2)}$, $S_n^{(3)}$, etc. Its influence in every additional sum is reduced by a factor of (1/3), $(1/3)^2$, $(1/3)^3$, ... respectively.

It is evident that the maximum error is

$$\sigma_{mx}(t) = 2\delta 3^{-q} + 2\delta 3^{-q+1} + \ldots + 2\delta 3^{-1} = \delta\left[1 - \left(\frac{1}{3}\right)^{q}\right]. \tag{4.37}$$

The maximum value of this error is when $q \to \infty$, and in this case $\sigma_{mx}(t) = \delta$. It is important to point out that in the case of processing the binary ΔM signal, the maximal error of multiplication is 2δ [2]. However, the maximum value of the multiplication constant for the binary case is 1, while for the ternary case is 0.5. More strictly, the total error of multiplication is the sum of error of quantization and the error introduced by the multiplier circuit i.e.

$$\xi(t) = \alpha \cdot \varepsilon_x(t) + \sigma(t) \tag{4.38}$$

where $\varepsilon_x(t)$ is the quantization error of TΔM. An example of FIR filter design is given in reference [7]. Fig. 4.5 shows an example of multiplication of sinusoidal input signal by the constant $\alpha = (0.43)_{10}$.

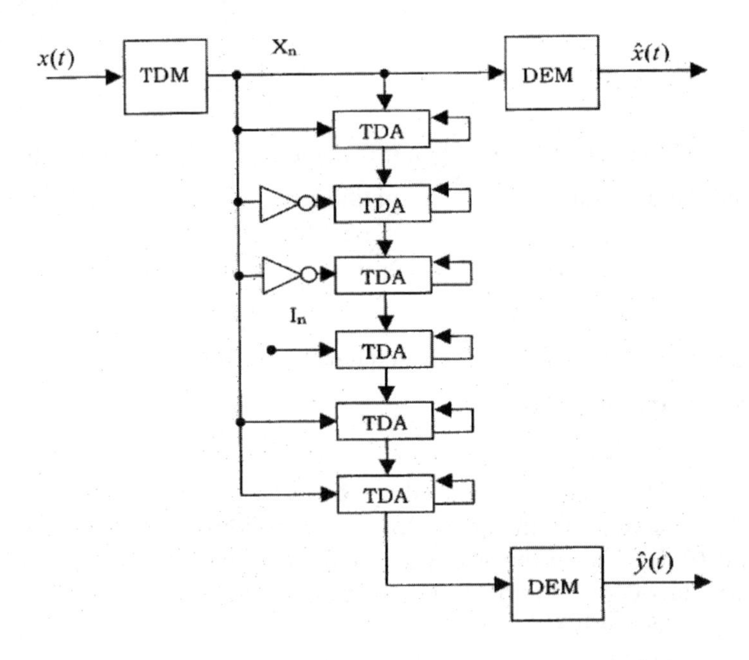


**Fig. 4.5.** Simulation block diagram of multiplication of TΔM signal $X_n$ with ternary constant α

The constant α can be represented in a ternary symmetrical system as $\alpha = (1\,1\,0\,-1\,-1\,1)_3$. The simulation results of the product are shown in Fig. 4.6.

It is possible to conclude, from Fig. 4.5, that the number of ternary adders used is equal to the number of the ternary digits needed to present the constant α. Zero value of a digit is presented with adder of idle input $I_n$. The serious problem of using binary [2] and ternary delta adder, as a building element of a multiplier, is its inherent attenuation property. Using ternary delta adder, it is possible to synthesize a multiplier to produce values less than 0.5. In addition, noise introduced by the delta adder can have serious consequences, in particular when adders are connected in series. Scaling-down properties of the binary delta adder are overcome by Kouvaras [8]. Here we will use a similar approach to solve the problem created by the scaling down of the ternary delta adder.

### 4.2.5 Ternary Delta Tripler

The inherent nature of attenuation of the delta adder presents a serious problem in implementation of digital filters [2]. When using the delta adder, a limited number of FIR and IIR filters can be synthesized. To solve this problem, a ternary tripler is introduced, Fig. 4.7 [4], [9].

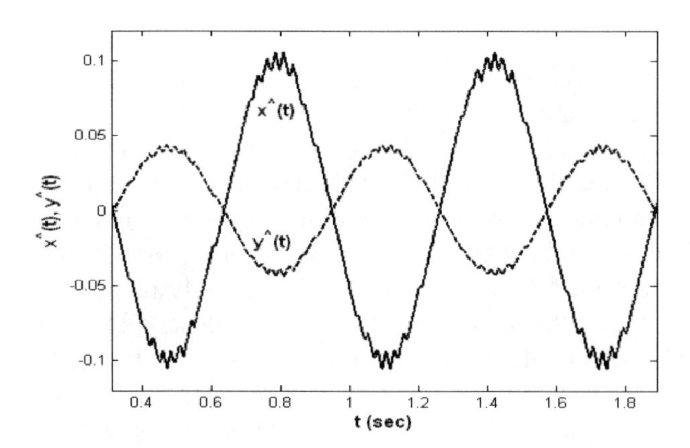


**Fig. 4.6.** Reconstructed signal $\hat{x}(t)$ (solid) and its scaled version $\hat{y}(t)$ (dashed)

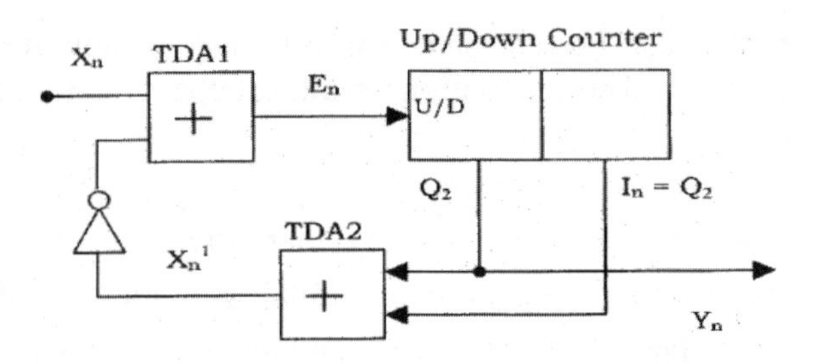


**Fig. 4.7.** Ternary tripler

The sequence $X_n$ is a ternary delta sequence which has to be multiplied by three to get TΔSQ $Y_n$. It consists of two ternary delta adders, and a ternary up-down counter. In the case when $E_n = +1$ ternary counter counts up, for $E_n = -1$ it counts down, and for $E_n = 0$ it remains unchanged. To operate an input signal, for which the maximum amplitude is (see chap. 1, eqn. 1.5),

$$V_{mx} \leq \left(\frac{\delta}{2\pi}\right)\left(\frac{f_s}{f_{in}}\right)$$

This is only a two-stage up-down counter, and second output ($Q_2$) represents an idle sequence $I_n$. Thus the output of $T\Delta A_2$ is

$$X_n^1 = \frac{1}{3} Y_n$$

This shows that $X_n^1$ is a TΔS of the one third of an analog signal corresponding to $Y_n$. By the use of a ternary delta adder TΔA1, $X_n^1$ is subtracted from $X_n$. It is evident that $E_n = +1$ as long as the analog value of $X_n$, $x(t)$, is greater than the analog value of $X_n^1$, $x'(t)$ (counter counts up). When $E_n = -1$, then the counter counts down, and for $E_n = 0$ counter remains unchanged. $E_n$ controls $Y_n$ at each clock pulse. In order for $X_n^1$ to be functioning properly it would have to emulate $X_n$ as closely as possible. In other words, for $Y_n = 3X_n$ the $X_n^1$ must follow $X_n$ very closely. In this case the error signal $E_n = 0$. Fig. 4.8a shows a simulation model. In Fig. 4.8b, results of simulation are shown when input signals are $s_1(t) = \sin\omega t$, and $s_2(t) = 0$. The number of samples per period was chosen in the simulation to be $N = 64$. It can be seen that the scaled signal of the sum $S_n$ is multiplied by three. An example of application of TΔM in synthesis of FIR filters can be found in reference [4]. Based on the presented results, we can conclude that arithmetic operations are possible not just on binary delta modulated signals, but on multi-valued TΔM signals as well. As an example, in the text that follows, we will show that arithmetic operations on a symmetric quaternary signals are possible as well.

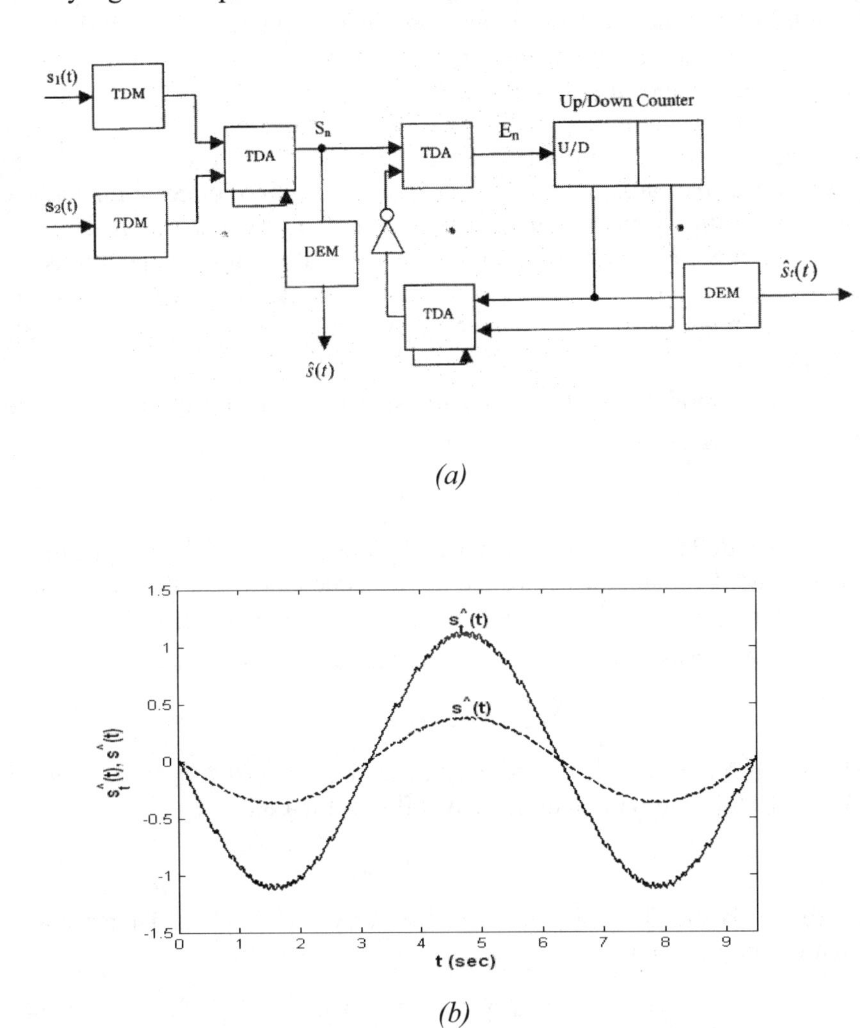


**Fig. 4.8.** (a) block diagram of simulation, (b) sum of reconstructed signals $\hat{s}(t)$ and its tripled version $\hat{s}_t(t)$

## 4.3 ADDITION OF MULTIVALUED TΔM SEQUENCES

In order to eliminate carry propagation chains, Avizienis [10] designed a so-called signed-digit number representation where the carry propagation during addition and subtraction is always limited to one position to the left. In a redundant symmetric radix four system, the signed-digit number is represented by a symmetrical 7-valued digit set {-3,-2, -1, 0, 1, 2, 3}. Since each digit in the radix four signed-digit number is no longer a quaternary digit, the ordinary quaternary circuits cannot be directly applied in the signed-digit number system. In order to show that arithmetic operations are possible on symmetrical quaternary ΔM signals, let us consider a quaternary delta modulator, which generates the signal $X_n \in \{-3,-2,-1,0,1,2,3\}$, Fig. 4.9.

The delta modulator of Fig. 4.9 transforms an analog input $x(t)$ to the quaternary sequence $\{X_n\} = \ldots X_{-1}, X_0, X_{+1}, \ldots$ Let

$$\{Y_n\} = \ldots, Y_{-1}, Y_0, Y_1, \ldots \tag{4.39}$$

be another delta sequence which will be considered as the output of an identical delta modulator controlled by the same clock. Let their integrated feedback output signals be $\hat{x}(t)$, $\hat{y}(t)$ and let the system errors be denoted by $\varepsilon_1(t)$, $\varepsilon_2(t)$, respectively. Let us define the quaternary sequence

$$\{S_n\} = \{X_n, Y_n\} \tag{4.40}$$

which will be termed the quaternary delta sequence (q.d.s) of the sum of $\{X_n\}$ and $\{Y_n\}$, and will be defined as follows. First let

$$Z_n = X_n + Y_n + C_{n-1} \tag{4.41}$$

where $C_{n-1}$ is the delayed version of the carry bit $C_n$. Consider table 4.3. Then we may check that

$$S_n = \frac{1}{2}\left(\frac{Z_n + 3.5}{|Z_n + 3.5|} + \frac{Z_n - 3.5}{|Z_n - 3.5|}\right) \tag{4.42}$$

$$C_n = Z_n - 3.5\left(\frac{Z_n + 3.5}{|Z_n + 3.5|} + \frac{Z_n - 3.5}{|Z_n - 3.5|}\right) \tag{4.43}$$

and

$$S_n = \frac{1}{7}\left(Z_n - C_n\right) = \frac{1}{7}\left(X_n + Y_n + C_{n-1} - C_n\right) \tag{4.44}$$

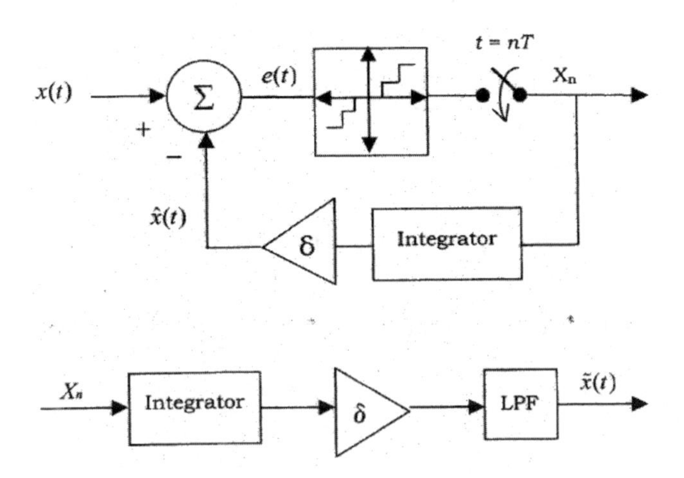


**Fig. 4.9.** Quaternary delta modulation systems

**Table 4.3.**

| $Z_n$ | $S_n$ | $C_n$ |
|---|---|---|
| 9 | 1 | 2 |
| 8 | 1 | 1 |
| 7 | 1 | 0 |
| 6 | 1 | -1 |
| 5 | 1 | -2 |
| 4 | 1 | -3 |
| 3 | 0 | 3 |
| 2 | 0 | 2 |
| 1 | 0 | 1 |
| 0 | 0 | 0 |
| -1 | 0 | -1 |
| -2 | 0 | -2 |
| -3 | 0 | -3 |
| -4 | -1 | 3 |
| -5 | -1 | 2 |
| -6 | -1 | 1 |
| -7 | -1 | 0 |
| -8 | -1 | -1 |
| -9 | -1 | -2 |

Using

$$\hat{x}(t) = T\delta \sum_{n=-\infty}^{N-1} X_n + \delta(t-NT)X_N \tag{4.45}$$

$$\hat{y}(t) = T\delta \sum_{n=-\infty}^{N-1} Y_n + \delta(t-NT)Y_N \tag{4.46}$$

and

$$\hat{s}(t) = T\delta \sum_{n=-\infty}^{N-1} S_n + \delta(t-NT)S_N \tag{4.47}$$

for $NT \le t \le (N+1)T$, we find that

$$T\delta \sum_{n=-\infty}^{N-1} S_n + \delta(t-NT)S_N \tag{4.48}$$

$$= \frac{T\delta}{7}\sum_{n=k}^{N-1}(X_n + Y_n + C_{n-1} - C_n) + \frac{\delta}{7}(t-NT)(X_N + Y_N + C_{N-1} - C_N$$

$$= \frac{T\delta}{7}\sum_{n=k}^{N-1} X_n + \frac{\delta}{7}(t-NT)X_N + \frac{T\delta}{7}\sum_{n=k}^{N-1} Y_n + \frac{\delta}{7}(t-NT)Y_N + \frac{\delta}{7}(t-NT)(C_{N-1} - C_N)$$

$$+ \frac{T\delta}{7}(C_{k-1} - C_{N-1})$$

Since $NT \le t \le (N+1)T$, we may let $t - NT = rT$, $0 \le r \le 1$. Then

$$+\frac{\delta}{7}(t-NT)(C_{N-1} - C_N) + \frac{T\delta}{7}(C_{k-1} - C_{N-1}) = \frac{T\delta}{7}[(r-1)C_{N-1} - rC_N + C_{k-1}] \tag{4.49}$$

Note that for $-3 \le C_i \le 3$ we have $-3 \le \{(r-1)C_{N-1} - rC_N\} \le 3$ and $-6 \le \{(r-1)C_{N-1} - rC_N + C_{k-1}\} \le 6$. Therefore,

$$\hat{s}(t) = \frac{1}{7}[\hat{x}(t) + \hat{y}(t)] + \Phi(t), \text{ where } |\Phi(t)| \le \frac{6T\delta}{7} \tag{4.50}$$

Note from Fig. 4.9 that $x(t) = \hat{x}(t) + \varepsilon_1(t)$ and $y(t) = \hat{y}(t) + \varepsilon_2(t)$ from which we have,

$$\hat{s}(t) = \frac{1}{7}\left[x(t) + y(t)\right] - \frac{1}{7}\left[\varepsilon_1(t) + \varepsilon_2(t)\right] + \Phi(t). \tag{4.51}$$

In eqn. (4.51), the expression $7^{-1}\left(\varepsilon_1(t)+\varepsilon_2(t)\right)$ is one-seventh the sum of the errors of the two ΔM systems and can be considered as an equivalent error of a ΔM system, the input of which is the analog signal $7^{-1}\left[x(t)+y(t)\right]$. We can see that the error $\Phi(t) \leq 6\delta T/7$ can be made small enough if the step size decreases while the sampling frequency correspondingly increases, such that $\delta T^{1}$ remains constant. It should be evident that an identical error bound holds for one seventh of the difference of two ΔM signals.

## 4.4 RESULTS OF SIMULATION

Some operations in digital signal processing are easily amenable to implementation with four-valued logic; for example, addition. Four-valued threshold logic full adder circuit implementations have been presented by Current and Mow [11]. Their presentation was not symmetrical.

The logical truth table of the symmetric four-valued full adder is given in table 4.4. We can see from table 4.3, eqns. (4.42) and (4.43), and table 4.4 that logical values for SUM and CARRY correspond to the arithmetical values $S_n$ and $C_n$ when SUM and CARRY are interchanged.

In fig. 4.10, we present a block diagram and the results of a computer simulation, where $x(t) = 2\sin\omega t$ and $y(t) = \sin(0.65\omega t)$. For this example, we have chosen $f_{\Delta M}/f_M = 1000$, and the smallest delta step size $\delta = 0.015$. From Fig. 4.10b, we see that the demodulated sum is really one-seventh of the actual sum plus an error. One part of the error is introduced by the quaternary delta full adder, and the other part is because of quantization.

In this book, we have shown that by the use of delta modulation in conjunction with symmetric quaternary logic it is possible to carry out the arithmetic operations of addition and subtraction, and by replication multiplication and division. If we had the hardware for a symmetrical quaternary shift register, and a quaternary delta modulator, it would be possible to synthesize a digital filter, which would have many of the advantages mentioned earlier [12].

**Table 4.4.**

| $C_{n-1}$ | $Y_n$ \ $X_n$ | S: -3 | -2 | -1 | 0 | 1 | 2 | 3 | C: -3 | -2 | -1 | 0 | 1 | 2 | 3 |
|---|---|---|---|---|---|---|---|---|---|---|---|---|---|---|---|
| | -3 | 1 | 2 | 3 | -3 | -2 | -1 | 0 | -1 | -1 | -1 | 0 | 0 | 0 | 0 |
| | -2 | 2 | 3 | -3 | -2 | -1 | 0 | 1 | -1 | -1 | 0 | 0 | 0 | 0 | 0 |
| | -1 | 3 | -3 | -2 | -1 | 0 | 1 | 2 | -1 | 0 | 0 | 0 | 0 | 0 | 0 |
| 0 | 0 | -3 | -2 | -1 | 0 | 1 | 2 | 3 | 0 | 0 | 0 | 0 | 0 | 0 | 0 |
| | 1 | -2 | -1 | 0 | 1 | 2 | 3 | -3 | 0 | 0 | 0 | 0 | 0 | 0 | 1 |
| | 2 | -1 | 0 | 1 | 2 | 3 | -3 | -2 | 0 | 0 | 0 | 0 | 0 | 1 | 1 |
| | 3 | 0 | 1 | 2 | 3 | -3 | -2 | -1 | 0 | 0 | 0 | 0 | 1 | 1 | 1 |
| | -3 | 2 | 3 | -3 | -2 | -1 | 0 | 1 | -1 | -1 | 0 | 0 | 0 | 0 | 0 |
| | -2 | 3 | -3 | -2 | -1 | 0 | 1 | 2 | -1 | 0 | 0 | 0 | 0 | 0 | 0 |
| | -1 | -3 | -2 | -1 | 0 | 1 | 2 | 3 | 0 | 0 | 0 | 0 | 0 | 0 | 0 |
| 1 | 0 | -2 | -1 | 0 | 1 | 2 | 3 | -3 | 0 | 0 | 0 | 0 | 0 | 0 | 1 |
| | 1 | -1 | 0 | 1 | 2 | 3 | -3 | -2 | 0 | 0 | 0 | 0 | 0 | 1 | 1 |
| | 2 | 0 | 1 | 2 | 3 | -3 | -2 | -1 | 0 | 0 | 0 | 0 | 1 | 1 | 1 |
| | 3 | 1 | 2 | 3 | -3 | -2 | -1 | 0 | 0 | 0 | 0 | 1 | 1 | 1 | 1 |
| | -3 | 0 | 1 | 2 | 3 | -3 | -2 | -1 | -1 | -1 | -1 | -1 | 0 | 0 | 0 |
| | -2 | 1 | 2 | 3 | -3 | -2 | -1 | 0 | -1 | -1 | -1 | 0 | 0 | 0 | 0 |
| | -1 | 2 | 3 | -3 | -2 | -1 | 0 | 1 | -1 | -1 | 0 | 0 | 0 | 0 | 0 |
| -1 | 0 | 3 | -3 | -2 | -1 | 0 | 1 | 2 | -1 | 0 | 0 | 0 | 0 | 0 | 0 |
| | 1 | -3 | -2 | -1 | 0 | 1 | 2 | 3 | 0 | 0 | 0 | 0 | 0 | 0 | 0 |
| | 2 | -2 | -1 | 0 | 1 | 2 | 3 | -3 | 0 | 0 | 0 | 0 | 0 | 0 | 1 |
| | 3 | -1 | 0 | 1 | 2 | 3 | -3 | -2 | 0 | 0 | 0 | 0 | 0 | 1 | 1 |

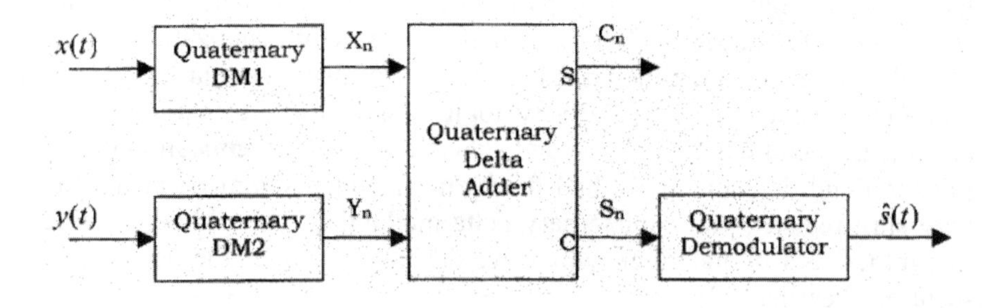


*(a)*

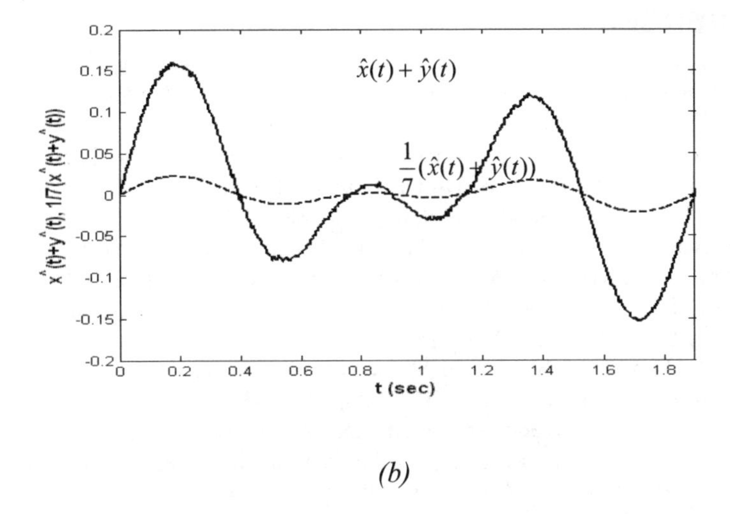


*(b)*

**Fig. 4.10.** (a) Block diagram of simulation model, (b) sum of reconstructed signals and their scaled demodulated version

## 4.5 CONCLUSION

In this chapter, a theory of arithmetic operations on multi-valued (ternary and quaternary) delta-modulated signals was derived. Error analysis of ternary delta adder and multiplier was done. In addition, the possibility of arithmetic operations on symmetric quaternary delta-modulated signals was shown. The main objective of this chapter was to show that, in addition to binary arithmetic operations, multi-valued arithmetic operations on the delta modulated pulse stream were possible as well.

## REFERENCES


1. L. Hurst, "The Logical Processing of Digital Signals," Crane Russak Co., New York, N.Y., 1978.
2. N. Kouvaros, "Operations on Delta-Modulated Signals and Their Applications in the Realization of Digital Filters," The Radio and Electronic Engineer, Sept. 1978, Vol. 48, No. 9, pp. 431-438.
3. Higuchi, T., Kameyama, M., "Ternary Logic System Based on a T-Gate," International Symposium on Multi Valued Logic, 1975, pp. 290-304.
4. Zrilic, Dj., "An Approach for Signal Processing Based on Use of Ternary Delta Modulation," Ph.D. Thesis, Beograd, Serbia, 1985.
5. R. W. Hamming, "Numerical Methods for Scientists and Engineers", Dover Publishers Inc., New York, ISBN 0-486-65341-6.
6. Dj. Zrilic, A. Mavretic, M. Freedman, "Arithmetic Ternary Operations on Delta-Modulated signals and their Applications in the Realization of Digital Filters" IEEE Tr. on ASSP, June 1985, pp. 760-764.
7. Zrilic, Dj., Mallinson, M., Zangi, K., Mavretic, A., Implamentation Signal Processing Functions on Ternary Encoded Delta-Modulated Pulse Streams, 1988 IEEE International Symposium on Circuits and Systems, Proceedings, Volume 2 of 3, pp. 1553-1556.
8. Kouvaros, N., A Special Purpose Delta Multiplier, The Radio and Electronic Engineer, Vol. 50, No. 4, September 1980, pp. 156-157.
9. Zrilic, Dj., A Special Purpose Ternary Delta Multiplier, Proceedings of the 30th Midwest Symposium on Circuits and Systems, Syracuse, NY, pp. 1355-1356.
10. Avizienis, A., Signed-Digit Number Representations for Fast Parallel Arithmetic, IRE Transactions on Electronic Computers, 1961, pp. 389-400.
11. K.W. Current and D.A. Mow, "Four-Valued Threshold Logic Full Adder Circuit Implementation," International Symposium on Multi-Valued Logic, pp. 95-100, 1978.
12. Dj. Zrilic, G. Kandus, M. Lozej, "Delta Modulation and its Applications in Digital Filtering," IASTED International Symposium, June 19-21, 1985, Paris, France.

# CHAPTER 5 NONLINEAR ARITHMETIC OPERATIONS

The objective of this chapter is to demonstrate how the Δ–ΣM format can be used to design many nonlinear functions of signal processing. For a given nonlinear function $y = f(x)$, we will show that it can take as input a discrete delta sequence (DΔS) $\{X_n\}$ corresponding to a signal $x(t)$ and produce DΔS $\{Y_n\}$, which on delta-demodulation yields $\hat{x}(t)$, a close reconstruction of $f(x(t))$. In this chapter we will closely follow the work of Freedman and Zrilic [7], and in addition we will present a number of novel simulation results.

## 5.1 BASIC Δ–ΣM CONCEPT

A delta sigma modulator (Δ–ΣM) is a device, which operates at a high frequency rate to convert an analog signal into a sequence $\{X_n\}$, $-\infty < n < \infty$ of binary bits. For convenience, we may view these bits as either +1 or -1. Such a binary one-bit sequence will be called a discrete delta sequence or DΔS. The block diagram of the delta sigma system is as in Fig. 5.1.

Let us take $f_s$ as the sampling rate and define $\Delta T = 1/f_s$. We then let $x_n = x(n\Delta T)$ represent the discretized input and let $\{X_n\}$ be the binary output sequence. The operation of our Δ–ΣM can be described by the equation

$$z_n = \sum_{j=-\infty}^{n-1} x_j \tag{5.1}$$

$$E_n = z_n - \delta \sum_{j=-\infty}^{n-1} X_j \tag{5.2}$$

or equivalently in recursive form

$$E_{n+1} = E_n + x_n - \delta X_n \tag{5.3}$$

where,

$$X_n = \operatorname{sgn} E_n = +1, \text{ for } E_n \geq 0,$$

$$X_n = \operatorname{sgn} E_n = -1, \text{ for } E_n < 0.$$

We thus view $\{X_n\}$ as the binary one-bit output sequence. In the linear Δ–ΣM, which we study in this book, $\delta$ will represent a fixed positive constant.

**Remark 1.1.** It is convenient to view $n$ as taking values $-\infty < n < \infty$, but in computer simulation and in theoretical work one would usually take $x_n = 0$ and $E_n = 0$ for $n < 0$. The Δ–ΣM will behave properly if the $\{E_n\}$ remains bounded for all $n$. In the next section we derive conditions for this to occur.

**Remark 1.2.** Associated with a Δ–ΣM, we also require a delta demodulator, which we denote by ΔDM. As we shall see, a delta demodulator is nothing more than an averaging filter. Precise error bounds for the operation of such a ΔDM will be given in the next section.

## 5.2 MATHEMATICAL PRELIMINARIES

We begin this section to prove some basic results about the operation of Δ–ΣM's.

**Lemma 2.1** *Let a Δ–ΣM be described for $n \geq 0$ by the equation*

$$E_{n+1} = E_n + x_n - \delta \operatorname{sgn} E_n \tag{5.4}$$

*with $E_0 = \alpha$. Assume that $|x_n| \leq \beta$ for all $n \geq 0$ and that $\beta \leq \delta$. Then*

*(i) If for some $n_1, |E_{n1}| \leq \beta + \delta$, then $|E_n| \leq \beta + \delta$ for all $n \geq n_1$.*

*(ii) If we further assume $\beta < \delta$ and $|A| > \beta + \delta$ then, letting*

$$n_* = \left\lceil \frac{|\alpha| - (\beta + \delta)}{\delta - \beta} \right\rceil,$$

*it will follow that $|E_n| \leq \beta + \delta$ for all $n \geq n_*$. (Here [ ] stands for the greatest integer function)*

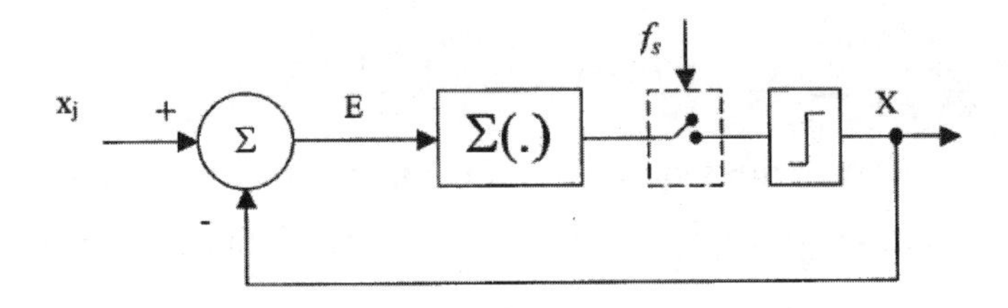


**Fig. 5.1.** Block diagram of Δ-ΣM system

**Proof:**

1. If $0 \le |E_{n1}| \le \beta + \delta$ then $x_{n1} - \delta \le E_n + x_{n1} - \delta \le \beta + \delta + x_{n1} - \delta$, so on using the fact that $|x_{n1}| \le \beta$ we get $-\beta - \delta \le E_{n1+1} \le 2\beta \le \beta + \delta$, i.e. $|E_{n1+1}| \le \beta + \delta$. On the other hand, if $-\beta - \delta \le E_{n1} < 0$ then $-\beta - \delta + x_{n1} + \delta \le E_n + x_{n1} + \delta < x_{n1} + \delta$, so that $-\beta - \delta < -2\beta \le E_{n1+1} \le \beta + \delta$, which again yields $|E_{n1+1}| \le \beta + \delta$, and the result follows by induction on $n$.
2. Without loss of generality, assume $E_0 = \alpha \ge 0$. Then inductively assume $E_0 \ldots En_{-1} \ge 0$. It follows that
   $0 \le E_1 \le \alpha + \beta - \delta$,
   $0 \le E_2 \le \alpha + 2(\beta - \delta)$
   ...
   $0 \le E_{n-1} \le \alpha + (n-1)(\beta - \delta)$.
   If $n_*$ is the smallest positive integer for which
   $$\alpha + n_*(\beta - \delta) \le \beta + \delta \tag{5.5}$$
   $|E_n| \le \beta + \delta$ for all $n \ge n_*$. It is clear that the value
   $$n_* = \left\lceil \frac{a - (\beta + \delta)}{\delta - \beta} \right\rceil$$
   will solve (5.5).

Property (ii) of Lemma 2.1 assures that no matter how large an $E_0$ we start with after $n_*$ time steps $E_n$ will lie in the appropriate range. This type of stability is important if our model is to accurately reflect a physical system. In the operation of a Δ–ΣM, a demodulator is required to recover the signal $x_n$ from the binary sequence $\{X_n\}$. This demodulator is generally an averaging filter. Let us now discuss the accuracy of such a filter.

**Definition 2.1.** Given the output $\{X_n\}_{-\infty}^{\infty}$ of a Δ–ΣM and an integer k>0, we describe, for each $n$, the reconstructed sequence $\{\hat{x}_n\}_{-\infty}^{\infty}$ via

$$\hat{x}_n = \frac{\delta}{2k+1} \sum_{j=-k}^{+k} X_{n+j} = \delta Ave_k(X_n) = \delta \hat{A}_\alpha^\beta(X_n),$$

where $\hat{A}_\alpha^\beta$ denotes the sequence averaging operator. It turns out that the reconstruction sequence $\hat{x}_n$ closely approximates the input sequence $x_n$.

**Theorem 2.1.** *If the averaging filter as above is used to demodulate a discrete delta sequence* $\{X_n\}$ *corresponding to* $x_n$, *then we have for all n*

$$|x_n - \hat{x}_n| \leq \frac{2(\beta+\delta)}{2k+1} + Sk(k+1)(\Delta T)^2 + O((k\Delta T)^4), \tag{5.6}$$

where

$$S = \max_{-\infty<t<\infty} |x''(t)|.$$

**Proof:** We assume $\{X_n\}$ is the output of a well functioning ΔΣM so that $\beta \leq \delta$ and $|E_n| \leq \beta + \delta$ for all $n$. In addition we assume that the continuous input $x(t)$ is a smooth function of $t$. From (5.3) we have for each $j$ $E_{n+j+1} = E_{n+j} + x_{n+j} - \delta X_{n+j}$, and summing both sides over $j$ from $j = -k$ to $j = +k$ gives

$$\sum_{j=-k}^{j=+k} E_{n+j+1} = \sum_{j=-k}^{j=+k} E_{n+j} + \sum_{j=-k}^{j=+k} x_{n+j} - \delta \sum_{j=-k}^{j=+k} X_{n+j} \tag{5.7}$$

so that,

$$E_{n+k+1} - E_{n-k} = \sum_{j=-k}^{j=+k} x_{n+j} - \delta \sum_{j=-k}^{j=+k} X_{n+j}. \tag{5.8}$$

Since $|E_n| \leq \beta + \delta$ for all $n$, it follows on dividing (5.8) by $2k+1$ and using Lemma 2.1(i),

$$\left| \frac{1}{2k+1} \sum_{j=-k}^{j=+k} x_{n+j} - \delta \frac{1}{2k+1} \sum_{j=-k}^{j=+k} X_{n+j} \right| \leq \frac{2(\beta+\delta)}{2k+1} \tag{5.9}$$

so that

$$\left| \frac{1}{2k+1} \sum_{j=-k}^{j=+k} x_{n+j} - \hat{x}_n \right| \leq \frac{2(\beta+\delta)}{2k+1} \tag{5.10}$$

To finish up we need to relate the expression on the left of the inequality (5.10) to $x_n$. It is clear that Theorem 2.1 follows by combining (5.10) with Lemma 2.2.

**Lemma 2.2** *Let $x(t)$ be a smooth function of t with* $S = \max_{-\infty<t<\infty} \left| x''(t) \right|$. *If k is a positive integer then*

$$\left| \frac{1}{2k+1} \sum_{j=-k}^{j=+k} x_{n+j} - x_n \right| \le Sk(k+1)(\Delta T)^2 + O((\Delta T)^4). \tag{5.11}$$

*Thus with the obvious notation we have*

$$\left| Ave_k(x_n) - x_n \right| \le Sk(k+1)(\Delta T)^2 + O((\Delta T)^4). \tag{5.12}$$

**Proof:** For the proof we diverge to calculus. Recalling Taylor's theorem of calculus,

$$x(t+h) = x(t) + x'(t)h + \frac{x''(t)h^2}{2!} + \frac{x'''(t)}{3!}h^3 + O(h^4)$$

it follows that with $h = j\Delta T$

$$x(t + j\Delta t) = x(t) + x'(t)j\Delta T + \frac{x''(t)}{2}(j\Delta T)^2 + \frac{x'''(t)}{6}(j\Delta T)^3 + O((j\Delta T)^4)$$

so that

$$\begin{aligned} &\frac{1}{2k+1} \sum_{j=-k}^{j=+k} x(t + j\Delta T) \\ &= x(t) + 0 + \frac{x''(t)}{2}(\Delta T)^2 \frac{1}{2k+1} \sum_{j=-k}^{j=+k} j^2 + 0 + O((\Delta T)^4) \\ &= x(t) + x''(t)(\Delta T)^2 \left( \frac{k(k+1)(2k+1)}{2k+1} \right) + O(\Delta T)^4 \\ &= x(t) + k(k+1)x''(t)(\Delta T)^2 + O(\Delta T)^4. \end{aligned}$$

It follows that for any $n$

$$\left| \frac{1}{2k+1} \sum_{j=-k}^{j=+k} x_{n+j} - x_n \right| \le Sk(k+1)(\Delta T)^2 + O((\Delta T)^4) \tag{5.13}$$

where $S = \max_{-\infty<t<\infty} \left| x''(t) \right|$ and this completes the proof.

**Remark 2.1.** For a DC-level input, $S=0$ and the result of Theorem 2.1 reduces to

$$|x_n - \hat{x}_n| \le \frac{2(\beta+\delta)}{2k+1}.$$

For a W-bandlimited signal with signal energy $\le E$, $\beta = (2E)^{1/2} < \delta$,

$$|\ S\ | = (32E)^{1/2} \times \pi^2 (k+1)\left(\frac{f_i}{f_s}\right)^2 + O(f_s^{-4}) \tag{5.14}$$

where $f_s$ and $f_i$ are sampling and input frequencies, respectively.

**Corollary 2.1.** *Suppose* $\{G_n\}_{n=0}^{\infty}$ *is a sequence of real numbers with* $|G_n| \le 1$. *Then the equations*

$$W_{n+1} = W_n + G_n - \operatorname{sgn} W_n, W_0 = \alpha, \text{ with } |\ \alpha\ | \le 2 \tag{5.15}$$

*define a sequence* $W_n$, *with* $|W_n| \le 2$. *We may define a new binary sequence* $\{Z_n\}_{n=0}^{\infty}$ *with* $Zn = \{+1, -1\}$ *via* $Z_n = \operatorname{sgn}(W_n)$. *It will follow that for any* $n>k$

$$|\ Ave_k(Z_n) - Ave_k(G_n)\ | \le \frac{4}{2k+1}. \tag{5.16}$$

*Thus, for k sufficiently large* $\{Z_n\}$ *and* $\{G_n\}$ *give 'equivalent' demodulations.*

**Remark 2.2.** A system in the form of (5.15) with $|W_n|$ bounded for all $n$ will be called *stable.*

**Remark 2.3.** In actual operation a ΔDM cannot anticipate the future. Given input $\{Z_n\}$ the output will be

$$DDM_l\{Z_n\} = \frac{\delta}{2l+1}\sum_{j=-2l}^{j=0} Z_{n+j} = \delta Ave_l\{Z_{n-l}\}.$$

Thus it will reproduce an input sequence $z(t)$ with delay time $l\Delta T$ to high accuracy. Letting $\Delta_l$ denote a delay of $l$ units, we have

$$DDM_l\{Z_n\} = \Delta_l \delta Ave_l\{Z_n\} = \delta Ave_l\{Z_{n-l}\}.$$

## 5.3 CONSTRUCTION OF NONLINEAR MEMORYLESS DEVICES

In this section we demonstrate how Δ–ΣM can be used to construct nonlinear memoryless devices. An example of a squarer will be given. In particular the problem is as follows. Given a DΔS $\{X_n\}_{n=-\infty}^{\infty}$ which corresponds to the output of a Δ–ΣM with input $x(t)$ we wish to construct a finite state machine that given $\{X_n\}$ as input produces an output DΔS. $\{Y_n\}$ which on delta-demodulation yields $\hat{y}(t)$ a close reconstruction of $f(x(t))$. To be more precise, assume we have a nonlinear real valued function $f$. Let us assume that $\max_{|x|\leq 1}|f(x)|\leq 1$ and that $f$ is continuous. For any integer $l \geq 2$, let $Q_l = \{a/l \mid a$ is an integer, $-l \leq a \leq l\}$. It is not difficult to show that $f$ may be arbitrarily closely approximated by maps in the form $\hat{f}: Q_l \to Q_L$ for $l$ and $L$ sufficiently large. For all practical purposes, we may assume that $f$ is exactly in the form of $\hat{f}$, i.e., we make the following assumptions.

**Assumptions 3.1.**
1. $f$ is smooth.
2. $\max_{|x|\leq 1}|f(x)|\leq 1$
3. $f$ restricted to $Q_l$, maps $Q_l$ to $Q_L$ for some appropriate $l$ and $L$.

Symbolically $f|_{Q_l}: Q_l \to Q_L$. If we now define for any $l \geq 2$, $Z_l = \{a \mid a$ is an integer with $-l \leq a \leq l\}$, we see that $f$ induces a map $F: Z_l \to Z_L$ via

$$F(a) = Lf\left(\frac{a}{l}\right).$$

Before proceeding let us give an example.

**Example 3.1.** Let $f(x) = x^2$ take $l=3$ and $L=9$. Then $f$ maps $Q_l$ to $Q_L$ and in fact $f(a/l) = a^2/l^2$ so that $F: Z_3 \to Z_9$ is given by $F(a) = a^2$ for $|a| \leq 3$.

**Definition 3.1.** Given $F: Z_l \to Z_L$, consider the finite state machine taking the DΔS sequence $\{X_n\}$ into the DΔS sequence $\{Y_n\}$ via the recursive scheme

$$W_{n+1} = W_n + F(X_n + \ldots + X_{n-l+1}) - L\,\mathrm{sgn}\,W_n,\; Y_n = \mathrm{sgn}\,W_n. \tag{5.17}$$

The recursive function described by (5.17) with input $\{X_n\}$ and output $\{Y_n\}$ will be denoted by $ALG_{l,L}(F)$. Thus $ALG_{l,L}(F)[X_n, \ldots, X_{n-l+1}] = Y_n$.

**Remark 3.1.** It is clear that at each stage the recursion (5.17) will produce a state $W_n$, which is an integer. We will show using Corollary 2.1 that $W_n$ can take on only a finite number of values and thus (5.17) describes a finite state machine with the possible $W_n$ values as its states. Given our nonlinear function *f*, the following theorem describes how $ALG_{l,L}(F)$ approximates $y(t)=f(x(t))$.

**Theorem 3.1.** *Let f(x) be a function defined on* [*0, 1*],

1. *f is smooth.*
2. $\max_{|x|\leq 1}|f(x)|\leq 1$
3. *f restricted to* $Q_{2s+1}$, *maps* $Q_{2s+1}$ *into* $Q_L$. *Let* $F : Z_{2s+1} \rightarrow Z_L$ *be the induced map given by*

$$F(a) = Lf\left(\frac{a}{2s+1}\right)$$

*for* $|a| \leq 2s+1$.

*Let x(t) be any smooth function of t with* $|x(t)|<1$ *and assume Δ–ΣM* $x(t)=\{X_n\}$. *Then the finite state machine* $ALG_{2s+1,L}(F)[X_n, \ldots, X_{n-2s}] = Y_n$ *produces a DΔS* $\{Y_n\}$ *with the property that for each positive integer k*

$$DDM_k\{Y_n\} \overset{def}{=} \hat{y}_n = \frac{\delta}{2k+1}\sum_{j=0}^{2k} Y_{n+j}$$

*and will satisfy*

$$|\hat{y}_n - f(x_{n-s-k})| \leq \frac{4}{2k+1} + \frac{2M(1+\delta)}{2s+1} \times (MSs(s+s)+k(k+1)N)(\Delta T)^2 \quad (5.18)$$

**Proof:** Recalling the definition of ALG(F) the recursive scheme defining $Y_n$ is as follows

$$W_{n+1} = W_n + F(X_n + \ldots + X_{n-l+1}) - L\,\mathrm{sgn}\, W_n, \quad (5.19)$$

$$Y_n = \mathrm{sgn}\, W_n. \quad (5.20)$$

Now each $X_l = \{{+1 \atop -1}\}$ so that the sum $X_n + \ldots + X_{n-2s} \in Z_{2s+1}$. By Assumption 3.1, (5.19) can be replaced by

$$W_{n+1} = W_n + Lf\left(\frac{X_n + \ldots + X_{n-2x}}{2s+1}\right) - \mathrm{sgn}\, W_n, \quad (5.21)$$

$$Y_n = \operatorname{sgn} W_n . \tag{5.22}$$

Note that all the elements of the $W_n$ sequence are integers. Defining $V_n = W_n/L$ and noticing that sgn $V_n$ = sgn $W_n$, (5.20) becomes equivalent to

$$W_{n+1} = W_n + Lf\left(\frac{X_n + ... + X_{n-2x}}{2s+1}\right) - \operatorname{sgn} W_n, \tag{5.23}$$

$$Y_n = \operatorname{sgn} W_n . \tag{5.24}$$

In the above

$$\left(\frac{X_n + ... + X_{n-2x}}{2s+1}\right) \in Q_{2s+1},$$

thus

$$f\left(\frac{X_n + ... + X_{n-2x}}{2s+1}\right)$$

is an element of $Q_L$ and so is certainly $\leq 1$ for each $n$. It follows from Corollary (2.1) that (5.21) and hence (5.20) is stable so that $V_n$ and/or $W_n$ can take on only a finite number of values and in addition

$$\left| Ave_k(Y_n) - Ave_k f\left(\frac{X_n + ... + X_{n-2x}}{2s+1}\right) \right| \leq \frac{4}{2k+1} \tag{5.25}$$

for all $s$ sufficiently large. Next, recalling the definition of

$$\hat{x}_n = \delta Ave_k(X_n) = \frac{\delta}{2k+1} \sum_{j=-k}^{+k} X_{n+j},$$

we have from Theorem 2.1.

$$|\hat{x}_n - x_n| \leq \frac{2(1+\delta)}{2s+1} + Ss(s+1)(\Delta T)^2 + O((k\Delta T)^4) \tag{5.26}$$

where

$$S = \max_{-\infty<t<\infty} |x^n(t)|$$

and so

$$\Delta_s \hat{x}_n = [\delta/(2s+1)](x_n + ... + x_{n-2s}) = \hat{x}_{n-s}$$

will satisfy

$$|\hat{x}_{n-s} - x_{n-s}| \leq \frac{2(1+\delta)}{2s+1} + Ss(s+1)(\Delta T)^2 \tag{5.27}$$

on neglecting $O((k\Delta T)^4)$ terms.

Let us set $M = \max_{|x|\leq 1}|f'(x)|$, then by the mean value theorem of calculus it follows that

$$|f(\hat{x}_{n-s}) - f(x_{n-s})| \leq M|\hat{x}_{n-s} - x_{n-s}| \tag{5.28}$$

so that combining (5.27) with (5.28) gives the estimate

$$|f(\hat{x}_{n-s}) - f(x_{n-s})| \leq \frac{2M(1+\delta)}{2s+1} + MSs(s+1)(\Delta T)^2 \tag{5.29}$$

again neglecting the $O((\Delta T)^4)$ terms.

If we now consider the difference between $Ave(f(\hat{x}_{n-s}))$ and $Ave(f(x_{n-s}))$, it is clear that the right-hand side of (5.29) will serve as an upper bound, i.e., we have

$$|\ Ave_k(f(\hat{x}_{n-s})) - Ave_k(f(x_{n-s}))\ | \leq \frac{2M(1+\delta)}{2s+1} + MSs(s+1)(\Delta T)^2 \tag{5.30}$$

Now using a little calculus we relate $Ave_k(f(x_{n-s}))$ to $(f(x_{n-s}))$. In fact using Lemma 2.2 on the smooth function $f(x(t))$ yields

$$|\ Ave_k(f(x_{n-s})) - (f(x_{n-s}))\ | \leq Nk(k+1)(\Delta T)^2 \tag{5.31}$$

where $N = \max_{-\infty<t<\infty}|(f(x(t)))''|$. Thus combining (5.30) and (5.31) gives us the estimate

$$\begin{aligned} &|\ Ave_k(f(\hat{x}_{n-s})) - Ave_k(f(x_{n-s}))\ | \\ &\leq \frac{2M(1+\delta)}{2s+1} + (MSs(s+1) + k(k+1)N)(\Delta T)^2 . \end{aligned} \tag{5.32}$$

Combining this with (5.25) now gives

$$\begin{aligned} &|\ Ave_k(Y_n) - f(x_{n-s})\ | \\ &\leq \frac{4}{2k+1} + \frac{2M(1+\delta)}{2s+1} + (MSs(s+1) + k(k+1)N)(\Delta T)^2 \end{aligned} \tag{5.33}$$

Lastly recalling that

$$DDM_l\{Y_n\} = \Delta_k Ave_k\{Y_n\} \overset{def\ n}{=} y_n, \tag{5.34}$$

we obtain the final estimate

$$| \hat{y}_n - f(x_{n-s-k}) | \leq \frac{4}{2k+1} + \frac{2M(1+\delta)}{2s+1} + (MSs(s+1) + k(k+1)N)(\Delta T)^2 \tag{5.35}$$

## 5.4 SOME SIMULATION RESULTS

First, let us understand how the algorithm works. For example, let us assume that we would like to perform squaring operations. Let the length of the memory register in fig. 5.2 be 100 bits, $l = 100$, and $X_n = \pm 1$ represents Δ-ΣM sequence. The content of the delay line is averaged first and then squared. Thus, the output of the detection logic circuit can be written as

$$F = \left( \sum_{i=n-100}^{n} X_i \right) \approx \left(100 \tilde{X}_n\right)^2$$
$$= 10,000 \overline{X}_n^2 .$$

According to fig. 5.2, the output $W_{n+1}$ can then be written as

$$W_{n+1} = F + W_n - L\,\mathrm{sgn}(W_n),\ or$$
$$W_n = F + W_{n-1} - L\,\mathrm{sgn}(W_{n-1}),$$
$$W_n \approx F + z^{-1} W_n - 10000 W_n$$
$$W_n \approx \frac{F}{1 - z^{-1} + 10000} = \frac{10000 \overline{X}_n^2}{1 - z^{-1} + 10000}$$
$$W_n \approx \frac{\overline{X}^2}{(1 - z^{-1})/10000 + 1} \approx \overline{X}^2 .$$

After demodulation of $\overline{X}^2$, analog signal $x^2(t)$ is obtained.

To multiply an analog signal $x(t)$ by some constant $\alpha$, the content of shift register of length $l$ has to be averaged and then multiplied by a constant $L$. For example, to multiply by constant $\alpha$=2, if $l$=100, then $L$=50. Thus,

$$W_n \approx \frac{100 \overline{X}_n}{1 - z^{-1} + 50} \approx \frac{2 \overline{X}_n}{(1 - z^{-1})/50 + 1} \approx 2 \overline{X}_n$$

After demodulation of $2\overline{X}_n$, analog signal $2x(t)$ is obtained.

### 5.4.1 Squaring Operation

As an example of a nonlinear operation on the Δ–ΣM pulse stream, let us take the squaring operation of Example 3.1. In our simulations, the input to the squaring circuit is a DΔS $X_n$ of a sine wave of a normalized frequency. The accuracy of the averaged (demodulated) signal depends on the length of the averaging $k$. The effect of R and $k$ on the accuracy of the reconstructed signal is well-known. To show the effect of register length $l$ on the squaring operation, $f(x) = x^2$, two different values of $l$ are chosen, $l = 80$ and $l = 100$. Input signal to be squared is of the form $x(t) = e^{-t}.sin(\omega t)$, where $f = 10$ Hz and sampling frequency $f_s = 1024$ Hz. A simulation model of the squaring operation is shown in Fig. 5.2. Fig. 5.3a shows the results of the squaring operation for $l = 80$. We see slight degradation of the squared signal. With $l = 100$ significant improvement is achieved.

There are a number of ways to realize the finite state machine of Fig. 5.2. Fig. 5.5 illustrates the state transition diagram, for example, when $l = 3$. In this case, we have to detect only $X_1$ and $X_3$ values in the detection logic circuit to get $F(X_1)=1$ and $F(X_3)=9$.

Finally, we would like to obtain a minimum value for the length of the averaging filter i.e., $k$ such that an output accuracy (after squaring) of 1% is achieved using the above-mentioned finite state machine. From (5.3) we have

$$E_{n+j+1} = E_{n+j} + \beta - X_{n+j}. \tag{5.36}$$

Summing (5.36) over $k$, we obtain

$$\sum_{j=0}^{k-1} E_{n+j+1} = \sum_{j=0}^{k-1} E_{n+j} + k\beta - \sum_{j=0}^{k-1} X_{n+j}. \tag{5.37}$$

This equation simplifies to

$$\frac{E_k - E_0}{k} = \beta - \frac{1}{k}\sum_{j=0}^{k-1} X_{n+j}, \tag{5.38}$$

where

$$\frac{1}{k}\sum_{j=0}^{k-1} X_{n+j} = Ave(X_n) = \hat{\beta}.$$

Because $|E_k| \leq 2$, , and assuming $E_0 = 0$, we have

$$|\beta - \hat{\beta}| \leq \frac{2}{k}. \tag{5.39}$$

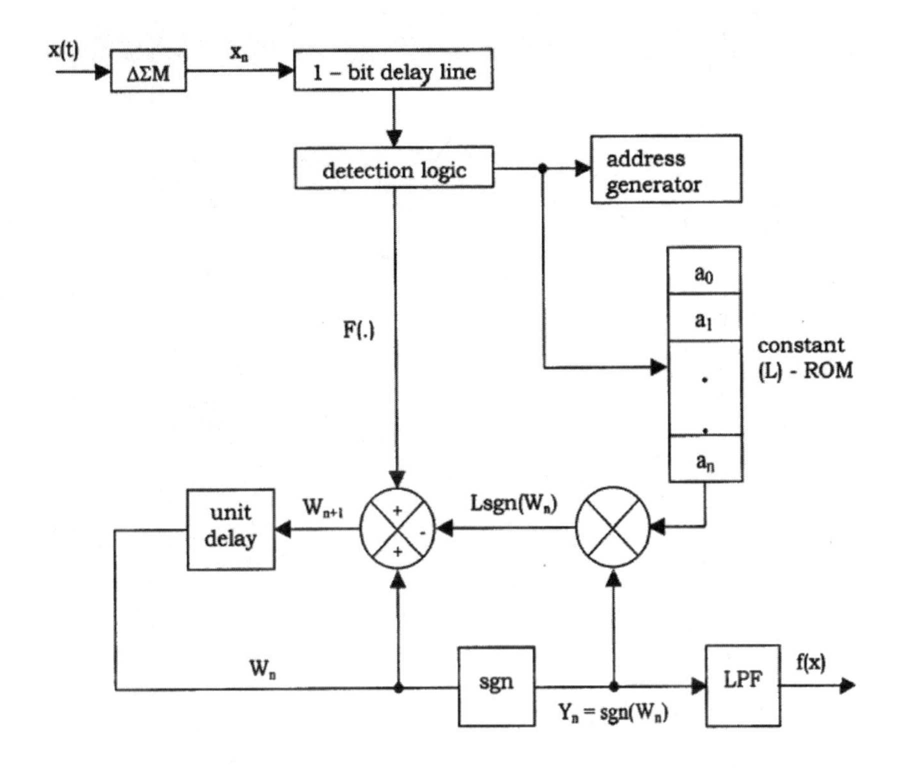


**Fig. 5.2.** Simulation model for the squaring operation

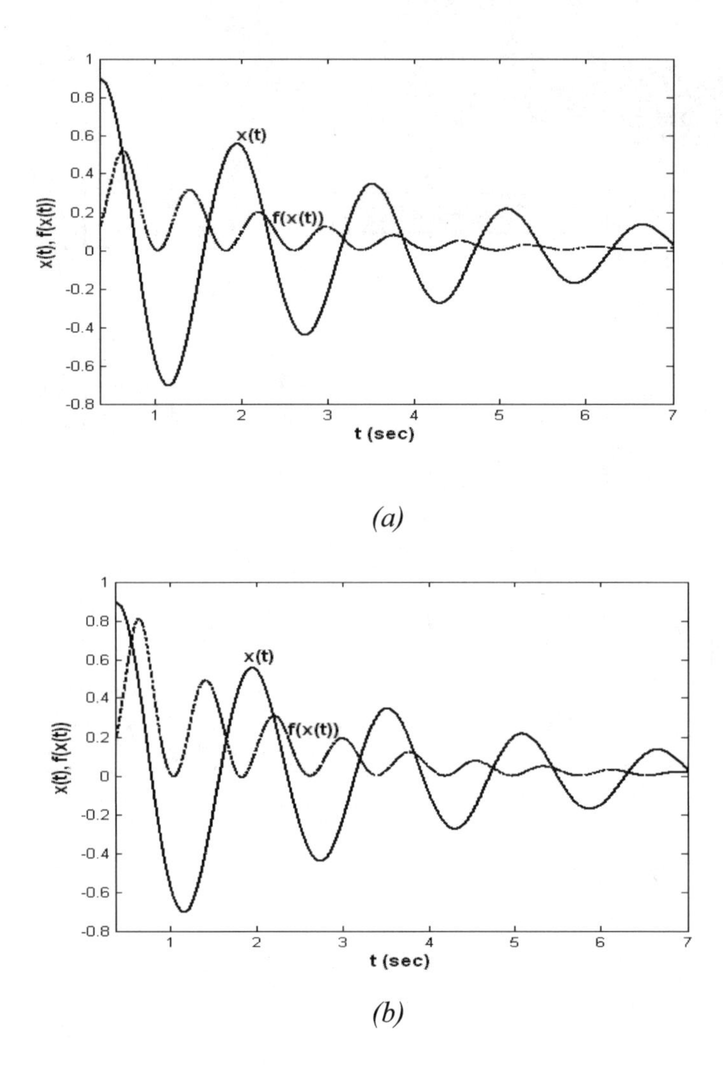


*(a)*

*(b)*

**Fig. 5.3.** Output of squaring operation, (a) with $l = 80$ and (b) $l = 100$

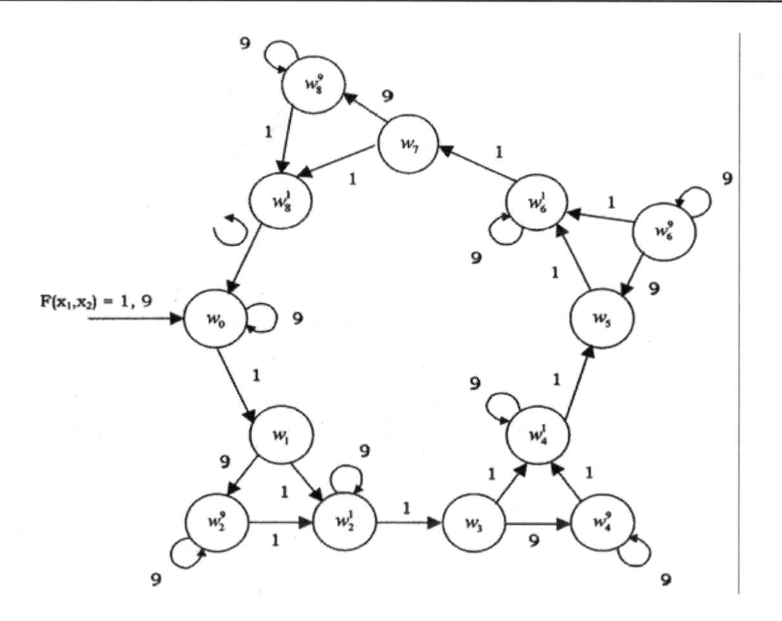


**Fig. 5.4.** State transition diagram, $l = 3$ [7]

If $k = 2^s$, then

$$| \beta - \hat{\beta} | \leq \frac{1}{2^{s-1}}. \tag{5.40}$$

The absolute value of the squarer output error is

$$\left|\beta^2 - \hat{\beta}^2\right| = \left|\beta + \hat{\beta}\right|\left|\beta - \hat{\beta}\right| \tag{5.41}$$

or

$$\left|\beta^2 - \hat{\beta}^2\right| \leq \frac{2}{2^{s-1}} = \frac{1}{2^{s-2}}. \tag{5.42}$$

To obtain better than 1% accuracy, $s \geq 9$ for $l \geq 31$, and R = 10000.

### 5.4.2 Mapping Of Boolean Functions

Using eqn. 5.17, we can recognize that it is possible to implement a state machine for binary mapping onto the algebraic domain. It is evident from eqn. 5.17 that as many logical outputs $Y_n$ as needed can be encoded in one equation. Thus, all we need to do is plug in the inputs to perform the arithmetic and simultaneously deduce all the outputs. Fig. 5.2 presents a general model of mapping Boolean functions onto non-logical domains .

It can be seen that, to perform binary to arithmetic function, the automaton requires a ΔM sequence as the input. The shift register with detection logic presents an averager of length $l$ in the case of multiplication of the input signal by a constant, or a squarer in the case of the squaring operation. This circuit can be implemented digitally using different approaches.

Fig. 5.5 presents an example of implementation of non-linear function,

$$z = \frac{x}{4}\left(1 - 4x^2\right).$$

Fig. 5.6 shows simulation results for $l = 100$ and over-sampling factor R = 2048. Curve I represents the theoretical value of the function $z(x)$, and curve II represents simulated value attenuated by a factor of two for the reason of comparison. In conclusion, we can state that by converting the analog input signal into the digital delta pulse stream, we are able to transform digital logic into arithmetic logic. The state-transition concept of the eqn. 5.17 is born from the automata theory.

### 5.4.3 Multiplication by A Constant Greater than One

As we have shown earlier, there exists an inherent problem of attenuation for both binary and multi-valued full-adder. The algorithm proposed in fig. 5.2 can be successfully applied for multiplication by a constant greater than one. Fig. 5.7 shows the case of multiplication by two. Delta half-adder has an attenuation of 0.5. The attenuated signal is fed into a finite-state machine. The length of the shift register is $l = 100$, and the value of constant $L = 50$. If $L = 25$, then signal $S_n$ is multiplied by four. In this example, input frequency is $f_m = 10$ Hz, and $f_s = 1024$ Hz.

### 5.4.4 Addition of Several Δ-ΣM Pulse Streams

The following example illustrates synthesis of a square-wave using four terms of the Fourier series,

$$V(t) = \frac{A}{2} + \sum_{n\ odd}^{\infty}\left(\frac{2A}{n\pi}\right)\sin 2\pi f_0 t\ .$$

In this example, $f_0$ is chosen to be 345 Hz, and $A = 1$V. Thus,

$$V(t) = \frac{1}{2} + \frac{2}{\pi}\sin 2\pi f_0 t + \frac{2}{3\pi}\sin 6\pi f_0 t + \frac{2}{5\pi}\sin 10\pi f_0 t + \frac{2}{7\pi}\sin 14\pi f_0 t$$

Fig. 5.8a presents a block diagram of simulation. As can be seen from the figure the square wave can be produced with addition of Δ-ΣM pulse sequences. An example of adding four terms from the series above is shown in fig. 5.8b. Four Δ-ΣM sequences $X_1$, $X_2$, $X_3$, and $X_4$ are added using universal Delta-Sigma Arithmetic Unit (DSAU) proposed in fig. 5.2.

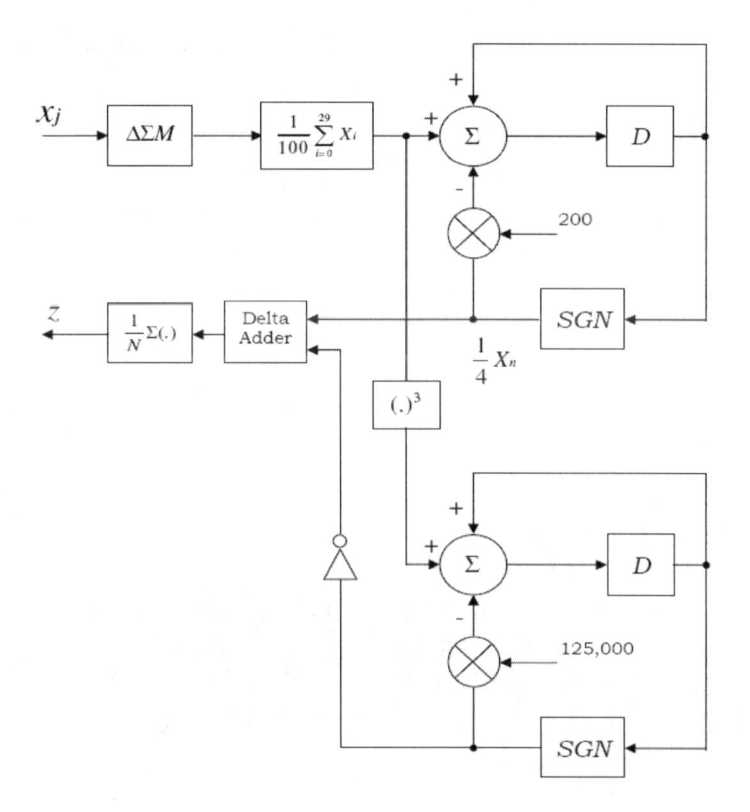


**Fig. 5.5.** Block diagram of realization of function *Z*(*x*)

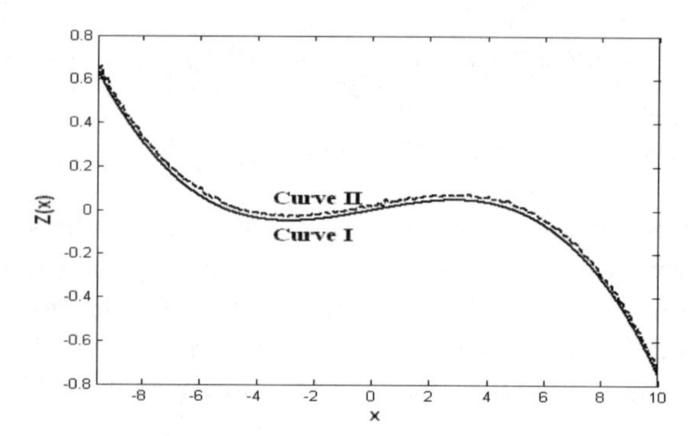


**Fig. 5.6.** Theoretical and simulation results of the non-linear function $Z = 0.25x - x^3$

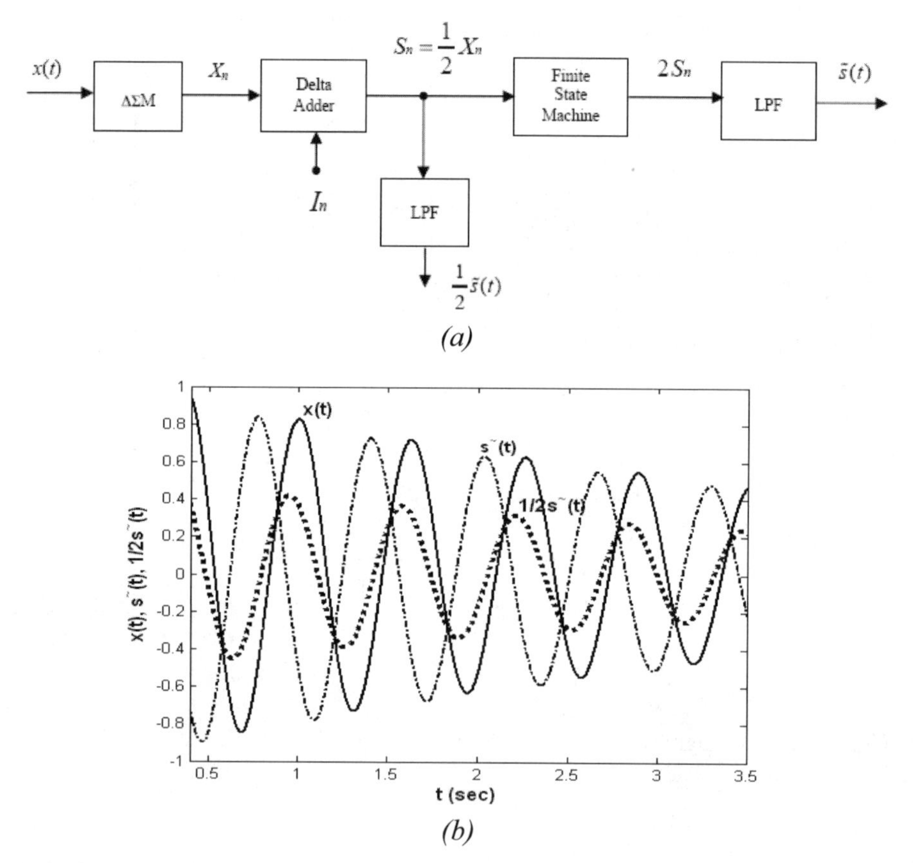


**Fig. 5.7.** (a) Block diagram of simulation and (b) respective waveforms

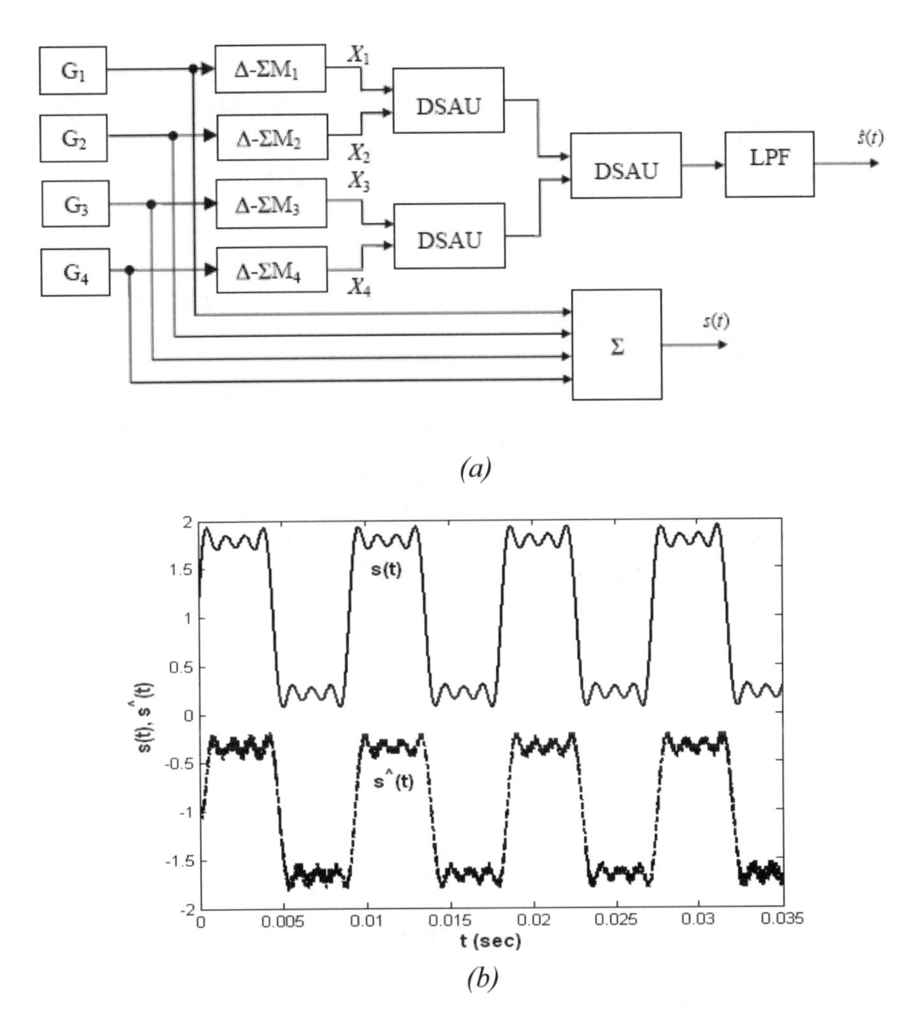


**Fig. 5.8.** (a) Block diagram of simulation, (b) theoretical and simulated output

## 5.5 CONCLUSION

In this chapter, theoretical developments of nonlinear operations on a delta modulated pulse stream were introduced. A universal algorithm for linear and nonlinear operation was presented and an error estimate was derived. A number of simulation examples were presented to demonstrate the possibility of linear and nonlinear operations on a delta-modulated pulse stream.

## REFERENCES


1. B.P. Agrawal and K. Shenoi, "Design Methodology for ΔΣM," *IEEE Trans. Commun.*, Vol. COM-31, No. 3, March 1983, pp. 360-370.
2. N. Kouvaras, "Operations on delta-modulated signals and their applications in the realization of digital filters", *The Radio and Electronic Engineer*, Vol. 48, No. 9, September 1978, pp. 431-438.
3. R.Steele, *Delta Modulation*, Wiley, New York, 1975.
4. D. Zrilic, M. Mallinson, K. Zangi, and A. Mavretic, "Implementing signal processing functions on ternary encoded delta-modulated pulse streams", *Proc. of IEEE ICAS '88*, Helsinki, pp. 1553-1556.
5. D. Zrilic, K. Zangi, A. Mavretic and M. Freedman, "Realization of digital filters for delta modulated signals", *Proc. of 30th Midwest Symposium on Circuits and Systems*, August 1987, Syracuse, N.Y., pp. 16-18.
6. D. Zrilic, G. Petrovic, B. Yuan, "Simplified Realization of Delta-Sigma Decoder", *Electronic Letters*, Vol. 33, No. 18, 1997, pp. 1515-1516.
7. M. Freedman, D. Zrilic, "Nonlinear Arithmetic Operations on Delta Sigma Pulse Stream", Signal Processing, Elsevier, 21 (1990), pp. 25-35.

# CHAPTER 6 MIXED PROCESSING OF Δ-ΣM SEQUENCES

## 6.1 INTRODUCTION

There have been several approaches to mixed mode processing of a Δ–ΣM pulse stream [3, 4]. In this chapter, we will follow the approach proposed in references [1, 2]. The specificity of a delta-modulated pulse stream is significant and in addition to direct linear and nonlinear processing, it can be used for mixed analog-digital processing as well. The objective of this chapter is to use the existence of the dual nature, both analog and digital, of a delta-modulated pulse stream in mixed mode signal processing. The dual nature of a Δ–ΣM pulse stream offers simple and cost effective solutions for many signal processing problems. In this chapter, we will consider the processing of a Δ–Σ modulated pulse stream, although the same results can be achieved using LΔM [1].

For an introduction to the process of Δ–Σ modulation under consideration, we will repeat some results from chap. 5. Let $x_n$ be an analog input signal sampled with frequency $f_s$, and define ΔT = $1/f_s$. Denote the Δ-Σ modulated signal by a sequence $X_n$, where each $X_n$ is either -1 or 1. This sequence is determined by a recursive relation as in [2]. The demodulated signal is obtained by low pass filtering of the modulated signal. We will consider a simple averaging filter of length $2k+1$

$$\hat{x}_n = \frac{\delta}{2k+1}\sum_{j=-k}^{k} X_{n+j}$$

where $\delta$ is a fixed positive constant.

Freedman and Zrilic showed [2] (Theorem 2.1) that

$$|x_n - \hat{x}_n| \leq \frac{4\delta}{2k+1} + Sk(k+1)(\Delta T)^2 + O\left((k\Delta T)^4\right)$$

where

$$S = \max_{-\infty<t<\infty} |x''(t)|.$$

The reader should be careful when perusing the original article that the $k$ in $O((k\Delta T)^4)$ was erroneously omitted. For our purpose, it is more useful to phrase the error bound in the following way:

$$|x_n - \hat{x}_n| \le \frac{2\delta}{k} + S(k+1)^2(\Delta T)^2 + O(k\Delta T)^4.$$

Freedman and Zrilic also consider the following idea (Section 3): Given two input streams $x$ and $y$, any reasonably smooth bounded function $f(x,y)$ may be approximated digitally after Δ–Σ modulation. The idea of doing so is elementary, simply demodulate $\hat{x}$ and $\hat{y}$. However, to do real time calculations, some care has to be taken to make the system causal. Approximate the domain with a mesh whose resolution is $1/(2L+1)$, and approximate the range with a mesh whose resolution is $1/k$. This will induce a function $F$ whose domain and range are integers. This function $F$ is used as the basis of a Δ–ΣM, and the resulting sequence is demodulated to approximate the correct result.

## 6.2 FURTHER RESULTS

Let us put these processes on a mathematical footing. First, suppose two signals, $x_n$ and $y_n$ are the inputs, but only one is Δ–Σ modulated. The modulated signal and the unmodulated signal are multiplied, and the resultant signal is then demodulated. Denote this modulated signal in the obvious way as $\overline{X_n y_n}$. Then for any $\in > 0$, with a sufficiently high sampling rate and a sufficiently long filter, we have that $\overline{X_n y_n}$ is a good approximation for $x_n y_n$, Fig. 6.1.

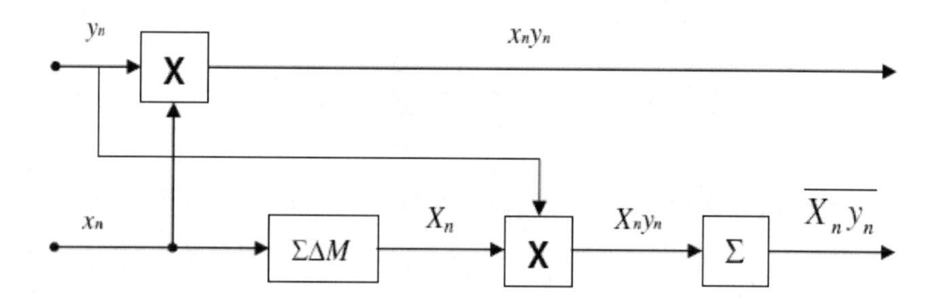


**Fig. 6.1.** Theorem 1 states that the output of these two circuits is virtually identical

To be precise let

$$R = \delta\left( \max_{\Delta T(n=k) \le t \le \Delta T(n \neq k)} y'(t) k \Delta T \right). \tag{6.1}$$

Let

$$M = \max\left(R, S^{1/2}, O^{1/4}\right) \tag{6.2}$$

and

$$Y = \max_t |y| \tag{6.3}$$

Then theorem 1 can be stated as below

**Theorem 1** *If*

$$k > \frac{4Y\delta}{\varepsilon}, \text{ and} \tag{6.4}$$

$$\Delta T < \frac{\varepsilon}{6M(k+1)} \tag{6.5}$$

*then*

$$\frac{\delta}{2k+1} \sum_{j=-k}^{k} y_{n+j} x_{n+j} - y_n x_n < \varepsilon \tag{6.6}$$

**Proof** Consider

$$\left| \frac{\delta}{2k+1} \sum_{j=-k}^{k} y_{n+j} x_{n+j} - y_n x_n \right| \tag{6.7}$$

$$\le$$

$$\left| \frac{\delta}{2k+1} \sum_{j=-k}^{k} y_{n+j} x_{n+j} - \frac{\delta}{2k+1} y_n \sum_{j=-k}^{k} x_{n+j} \right| + \left| \frac{\delta}{2k+1} y_n \sum_{j=-k}^{k} x_{n+j} - y \right.$$

$$\le \quad \left| \frac{\delta}{2k+1} \sum_{j=-k}^{k} \left(y_n - y_{n+j}\right) x_{n+j} \right| + |y_n| |\hat{x}_n - x_n|$$

$$\leq \quad \delta\left(\frac{2k+1}{2k+1}\max_{n-k\leq j\leq n+k}\left|y_n - y_{n+j}\right|\right)+\left|y_n\right|\left|\hat{x}_n - x_n\right|$$

$$\leq \quad \delta\left(\max_{\Delta T(n-k)\leq j\leq \Delta T(n+k)} y'(t)k\Delta T\right)+\left|y_n\right|\left|\hat{x}_n - x_n\right|$$

$$= \quad R+\left|y_n\right|\left|\hat{x}_n - x_n\right|$$

$$\leq \quad R+\left|y_n\right|\left|\frac{2\delta}{k}+S(k+1)^2\Delta T^2+c(k\Delta T)^4\right|$$

$$\leq \quad R+\left|\frac{Y2\delta}{k}\right|+\left|YS(k+1)^2\Delta T^2\right|+\left|Yc(k\Delta T)^4\right|$$

$$< \quad \frac{\varepsilon}{6}+\frac{\varepsilon}{2}+\frac{\varepsilon}{6}+\frac{\varepsilon}{6}$$

$$= \quad \varepsilon .$$

This theorem is noteworthy on its own, although it is superseded by the following result. Consider a bounded, well-behaved function of two inputs $f(x,y)$. The spirit of the idea is to approximate the domain by multiples of $1/(2L + 1)$ and the range by multiples of $1/k$. Ideally, we would have the following situation

$$E_{n+1} = E_n + f(x_n, y_n) - \operatorname{sgn} E_n \tag{6.8}$$

but this will not do, since we have no control over the domain and range of $f$. We can, however, approximate $x_n$ closely by

$$\hat{x} = \frac{\delta}{2L_x + 1}\sum_{j=-L}^{L} X_{n-L+j} \tag{6.9}$$

and similarly for $y_n$. This is actually a close approximation to $x_{n-k}$, not $x_n$, but

$$\left|x_n - x_{n-k}\right| < P_x k\Delta T \tag{6.10}$$

where

$$P_x = \max_t \left|x'\right| .$$

These approximations are automatically multiples of $1/(2L_x + 1)$. If we approximate the range of $f$ by multiples of $1/k$, we would have

$$E_{n+1} = E_n + \left(f\left(\tilde{x}_n, \tilde{y}_n\right) + \delta_n\right) - \operatorname{sgn} E_n = E_n + \left(f\left(\tilde{x}_n, \tilde{y}_n\right) + \delta_n\right) - F_n \quad (6.11)$$

where $|\delta_n| \leq 1/2k < 1/k$, and $F_n = \operatorname{sgn} E_n$. In the natural way, we define

$$\tilde{f}_n = \left(f\left(\tilde{x}_n, \tilde{y}_n\right) + \delta_n\right) \quad (6.12)$$

and

$$\hat{f} = \frac{\delta}{2k+1} \sum_{j=-k}^{k} F_{n-k+j} \quad (6.13)$$

$$\left|f(\tilde{x}_n, \tilde{y}_n) - \hat{f}_n\right| \leq \frac{1}{k}. \quad (6.14)$$

The point being made here is that $\hat{f}_n$ is a reconstruction of $\tilde{f}_n$, which is within $1/k$ of $f\left(\tilde{x}_n, \tilde{y}_n\right)$, and this is quite close to $f\left(x_n, y_n\right)$. Let us make that all precise, let

$$D^y = \max \frac{\partial f}{\partial y}, \; D^x = \max \frac{\partial f}{\partial x}. \quad (6.15)$$

We now give the following theorem.

**Theorem 2** *For $\varepsilon > 0$, $k$, $L_x$, $L_y$, and $\Delta T$ may be chosen so that*

$$\left|f\left(x_n, y_n\right) - \hat{f}_n\right| < \varepsilon, \quad (6.16)$$

(See Fig. 6.2)

**Proof** Choose $k$, $L_x$, $L_y$, and $\Delta T$ so that, $\left|x_n - \tilde{x}_n\right| < \varepsilon/\left(3D^x\right)$, $\left|y_n - \tilde{y}_n\right| < \varepsilon/\left(3D^y\right)$, and $\left|\tilde{f}_n - \hat{f}_n\right| + 1/k < \varepsilon/3$. Then,

$$\left|f\left(x_n, y_n\right) - \hat{f}_n\right| \leq \left|f\left(x_n, y_n\right) - \tilde{f}_n\right| + \left|\tilde{f}_n - \hat{f}_n\right| \quad (6.17)$$

$$\leq \left|f\left(x_n, y_n\right) - f\left(\tilde{x}_n, \tilde{y}_n\right)\right| + \frac{1}{k} + \left|\tilde{f}_n - \hat{f}_n\right|$$

$$\leq \left| f(x_n, y_n) - f(\tilde{x}_n, \tilde{y}_n) \right| + \frac{\varepsilon}{3}$$

$$\leq \left| f(x_n, y_n) - f(x_n, \tilde{y}_n) \right| + \left| f(x_n, \tilde{y}_n) - f(\tilde{x}_n, \tilde{y}_n) \right| + \frac{\varepsilon}{3}$$

$$\leq D^y \left| y_n - \tilde{y}_n \right| + D^x \left| x_n - \tilde{x}_n \right| + \frac{\varepsilon}{3}$$

$$< \frac{\varepsilon}{3} + \frac{\varepsilon}{3} + \frac{\varepsilon}{3}$$

$$= \varepsilon .$$

## 6.3 OPTIMIZATION

In the proofs above, we have repeatedly used the identity

$$\left| x_n - \hat{x}_n \right| \leq \frac{4\delta}{2k+1} + Sk(k+1)(\Delta T)^2 + O\left((k\Delta T)^4\right). \tag{6.18}$$

We now turn our attention to optimizing the parameters $\Delta T$ and $k$. To begin, we make the simplifying assumption that the $O((k\Delta T)^4)$ term is ludicrously small and may be safely ignored. By asymptotic expansions, this

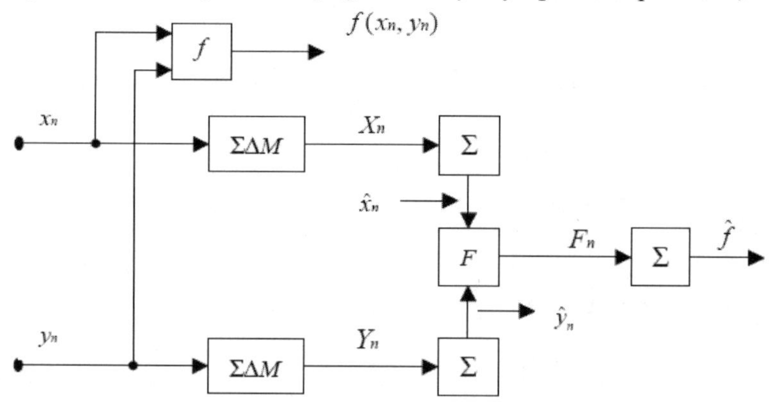


**Fig. 6.2.** Theorem 2 states that the output of these two circuits is virtually identical

turns out to be roughly equivalent to assuming that $S/\delta \leq f_S^4$. For example, if $f_s = 10^2$, then $S/\delta < 10^8$. If these were false, the system would be highly under-sampled.

At any rate, since it never hurts to highly over-sample a system, we'd like to fix $\Delta T$ at its maximum and optimize $k$. To do so, we'll minimize the error expression with respect to $k$. If

$$E(k) = \frac{4\delta}{2k+1} + Sk(k+1)(\Delta T)^2, \tag{6.19}$$

then

$$E'(k) = \frac{-8\delta}{(2k+1)^2} + S(2k+1)(\Delta T)^2. \tag{6.20}$$

E has a minimum if and only if $E'(k) = 0$ and,

$$E''(k) = \frac{8\delta}{(2k+1)^3} + 2S(\Delta T)^2 > 0 \tag{6.21}$$

which is certainly true. We have reduced the problem to the following calculation

$$\frac{-8\delta}{(2k+1)^2} + S(2k+1)(\Delta T)^2 = 0 \ldots \tag{6.22}$$

$$S(2k+1)(\Delta T)^2 = \frac{8\delta}{(2k+1)^2} \ldots$$

$$(2k+1)^3 = \frac{8\delta}{S(\Delta T)^2} \ldots$$

$$(2k+1) = \left(\frac{8\delta}{S(\Delta T)^2}\right)^{1/3} \ldots$$

$$k = \frac{2}{2}\left(\frac{\delta}{S(\Delta T)^2}\right)^{1/3} - \frac{1}{2} \ldots$$

$$k = \left(\frac{\delta}{S(\Delta T)^2}\right)^{1/3} - \frac{1}{2}$$

This is the optimum value for $k$ with a given $f_s$. Of course, this estimate is slightly inaccurate, since the error estimate is not exact, so some experimentation should be done to compare this theoretically optimal value with a practical evaluation. Fig. 6.3 shows $k$ as a function of sampling frequency and Fig. 6.4 shows the value of error as a function of $k$.

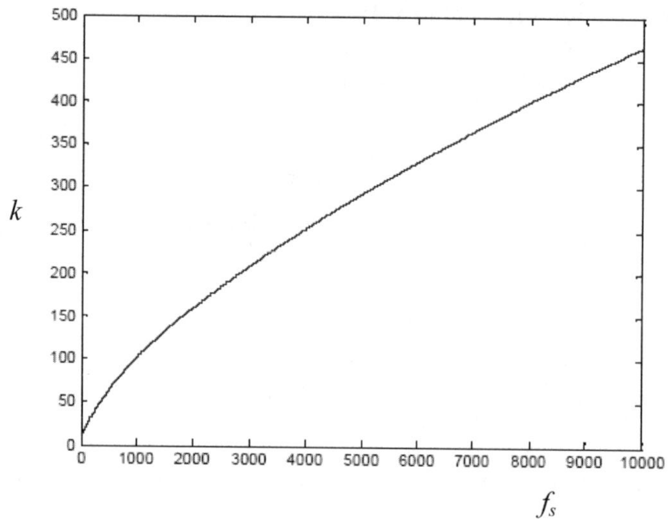


**Fig. 6.3.** $k$ as a function of $f_s$

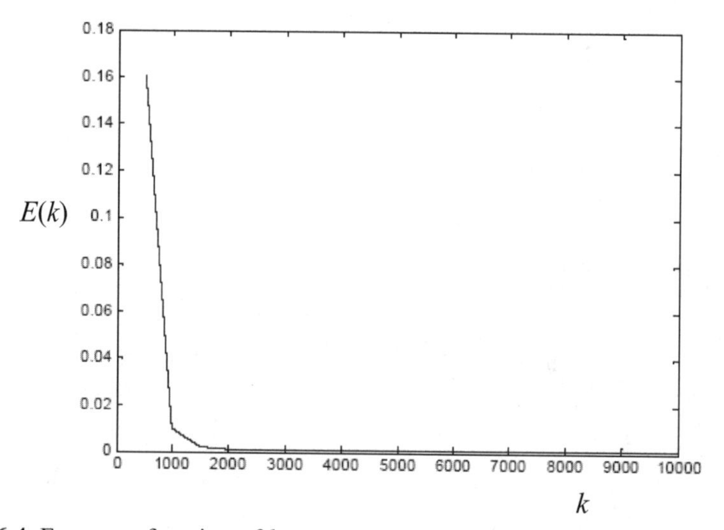


**Fig. 6.4.** Error as a function of $k$

## 6.4 SOME SIMULATION RESULTS

### 6.4.1 Mixed Mode Multiplication

To investigate the behavior of the proposed circuit from Fig. 6.1, simulations were performed. We assume a first-order delta-sigma modulator clocked at a rate of $f_s$ = 2 kHz. The input was a sinusoidal signal of frequency $f_{in}$ = 3 Hz. The averager (demodulator) was chosen to be a sixth order Butterworth low-pass filter. Fig. 6.5 shows the simulation block diagram. Relevant waveforms are shown in fig. 6.6. We can see that the output signal is double the frequency of the input. Consequently, the system in fig. 6.5 functions like a frequency doubler. In this example, the input signal is $e^{-t}\sin\omega t$.

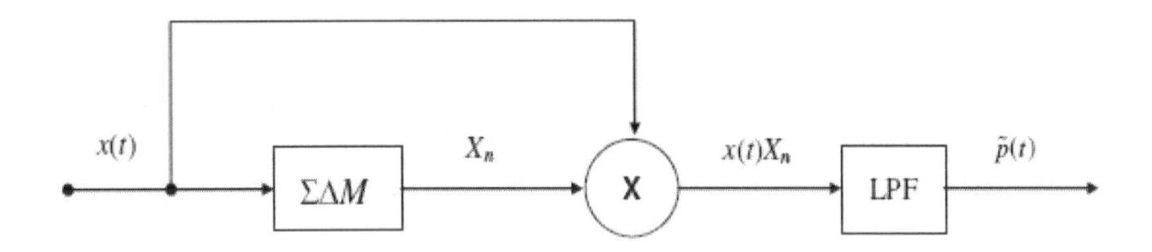


**Fig. 6.5.** Simulation block diagram

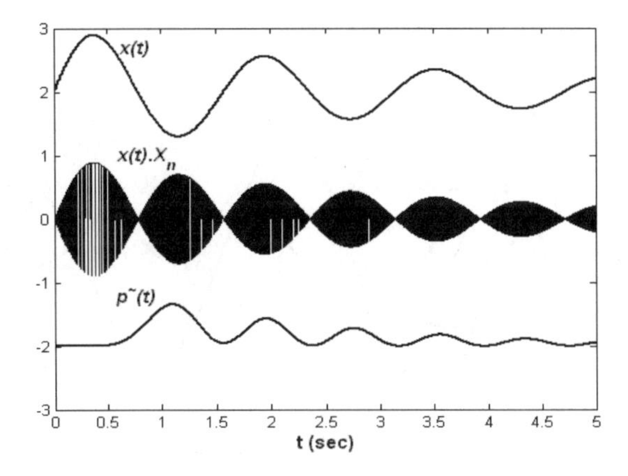


**Fig. 6.6.** Relevant waveforms from fig. 6.5

As a second example, a frequency doubling operation is shown in fig. 6.7. The difference is that amplitude scaling is done using a digital delta half-adder, which introduces attenuation by ½. Fig. 6.8 shows the corresponding input and output waveforms. It can be seen that frequency doubling is achieved by multiplication of an analog input signal with its digital pulse stream.

It is worth noting that the output signal is scaled by a factor of 2 and that it is slightly degraded; this is due to the introduction of a Delta-adder (DA). This problem can be solved by increasing sampling frequency, or length of the averaging filter, or both. Our simulations have shown that the mixing approach may be applied to the modified circuit in many different cases. Of particular importance are the cases when it is necessary not only to process a Δ–Σ pulse stream, but an analog signal as well.

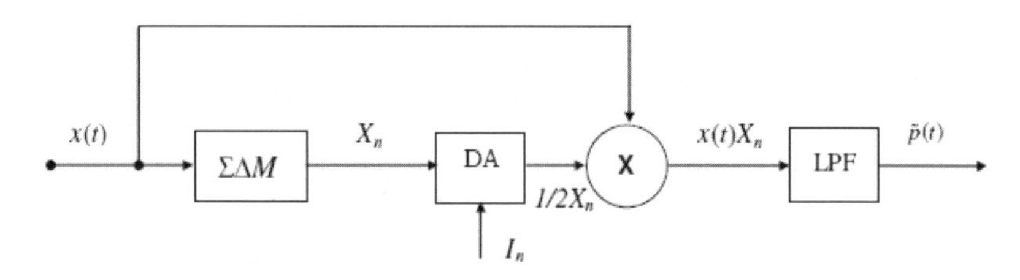


**Fig. 6.7.** Simulation block diagram when scaling with a constant a = 0.5 is required

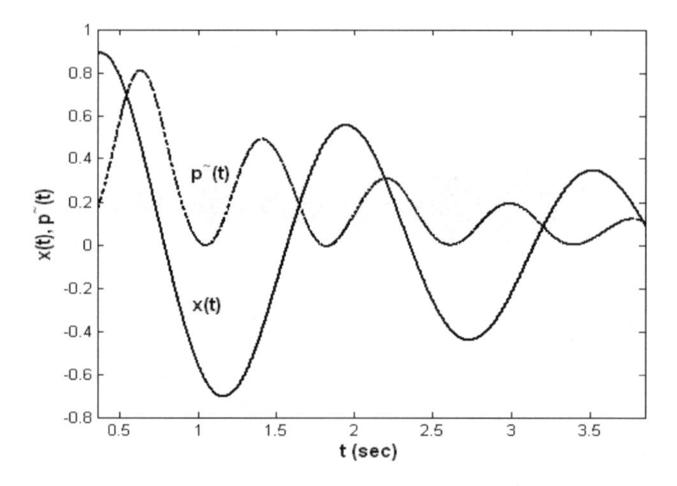


**Fig. 6.8.** Relevant waveforms of the signals at different points in the block diagram of fig. 6.7

## 6.4.2 Mixed Mode Multiplication of an Arbitrary Analog Signal

An example of multiplication of a Δ–Σ pulse stream with an arbitrary analog input signal is shown in Fig. 6.9. The results of this simulation are shown in Fig. 6.10, where $y(t) = e^{-t}$ and $x(t) = \sin\omega t$. Depending on the application, it is evident that numerous functions can be realized using mixed analog/digital processing based on Δ-ΣM. Fig. 6.11(a) shows an example of mixed processing

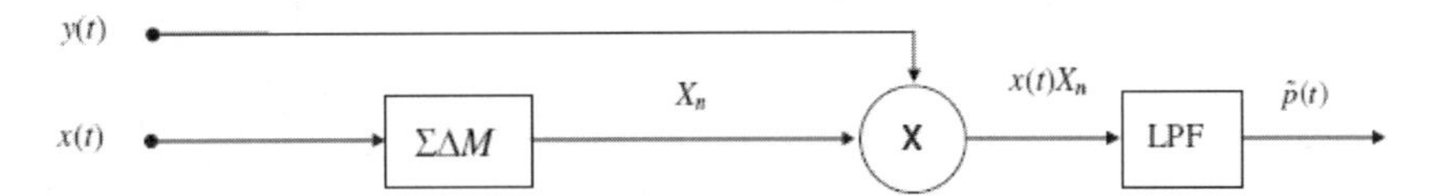


**Fig. 6.9.** Simulation block diagram

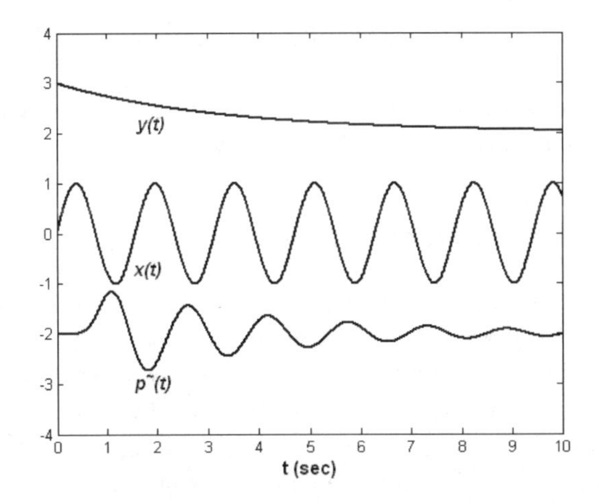


**Fig. 6.10.** Relevant waveforms from the simulation block diagram of fig. 6.9

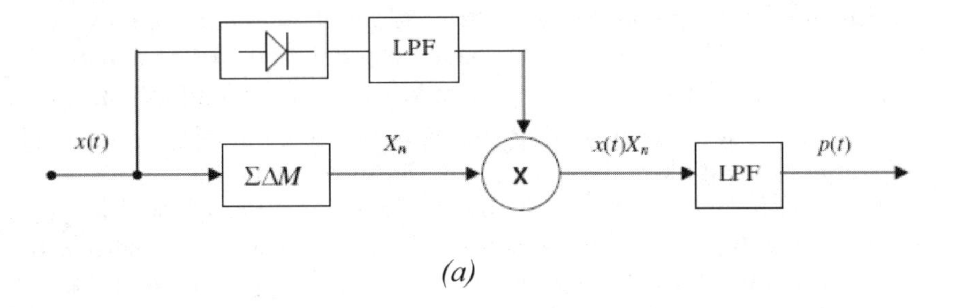


*(a)*

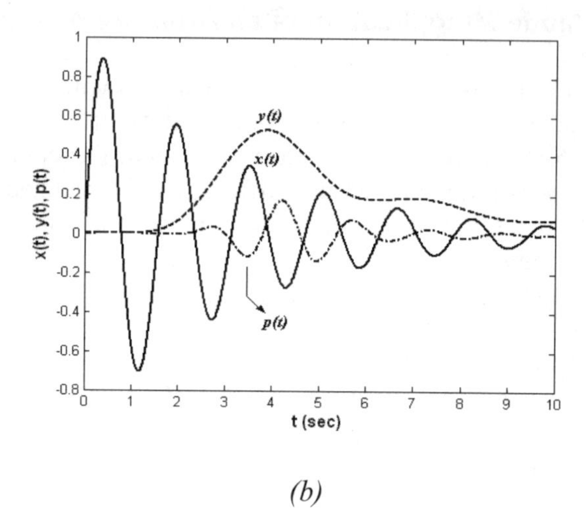


*(b)*

**Fig. 6.11.** (a) Simulation block diagram, and (b) relevant waveforms

where the input signal is rectified, low-pass filtered and then multiplied by Δ-Σ pulse stream of the original analog input signal.

### 6.4.3 A Robust Amplitude Modulation System

According to the well known equation for AM waveforms, $e(t) = (E_c + E_m\cos\omega_m t)\cos\Omega_c t$, the most critical element to be implemented is a circuit for multiplication. It is traditionally called a mixer, and its role is to translate a low-frequency signal of frequency $\omega_m$ to some higher frequency range, $\Omega_c \pm \omega_m$. In fig. 6.12 (a) a novel AM system is presented based on mixed analog /digital processing. From the proposed block diagram we can see that a conventional multiplier is replaced by an analog multiplexer, whose control input is $C_n$. It is important to point out that low frequency information of the signal $x(t)$ is contained in both amplitude and carrier of the signal $e(t)$. Thus, the asynchronous demodulation can be achieved in two ways. The signal $\hat{x}(t)$ is obtained when identical multiplying multiplexer is used at the transmitting side and the receiving side as well. The comparator circuit plays the role of a carrier recovery circuit.

We demonstrate the principle by implementing the proposed system shown in fig. 6.12 (a). This system is implemented with inexpensive off-the-shelf components. Analog Devices *ADMOD79*JQ* is used and a simple two-channel analog multiplexer is implemented with the *CMOS CD4066* bilateral switch.

The AM demodulator consists of a zero-crossing detector, multiplexer (the same as on transmitting side), and a low-pass filter. It is important to point out that low-frequency information of the signal $x(t)$ is contained in both the envelope and the carrier of $e(t)$. The zero-crossing detector is used to detect the carrier signal $C_n$. After multiplexing and filtering the signal, $\hat{x}(t)$ is received. Fig. 6.12 (b) shows simulation waveforms of the AM system from fig. 6.12 (a), with a sinusoidal input of frequency of 10 Hz and a sampling frequency of 1 kHz. Fig. 6.12 (c) shows experimental waveforms, and fig. 6.12 (d) shows relevant waveforms when Gaussian noise is added. It is clear that the system behaves properly as long as the amplitude of the noise does not cross the threshold of the detector. As a low-pass filter an integrator of the first order is used. Fig. 6.13 presents a synchronous AM system, where the carrier signal is inserted into a Δ-ΣM pulse stream using a Manchester encoder. The system is simulated in the presence of Gaussian noise. Fig. 6.13 (c) shows the case where the modulated signal is totally corrupted by noise. Fig. 6.13 (d) shows waveform of the received signal $\hat{x}(t)$. Figs. 6.13 (e) and (f) show Manchester encoded and decoded signals $C_n$ and $D_n$, respectively, when Gaussian noise is added into channel. Fig. 6.13 (g) shows frequency spectrum of the AM modulated waveform. The advantage of our approach is simple and inexpensive implementation and low power consumption of the system. The system can be implemented on a single VLSI chip. In addition, the dual nature of information content (in the envelope and in the carrier) of the AM signal $e(t)$ has a significant benefit in the presence of channel noise.

Fig. 6.14 presents one possible implementation of the multiplexing multiplier [1]. There are applications where precise splitting and control of analog waveforms in digital form is required. Usually rectification is done by diodes, which has its advantages such as low size and cost. Unfortunately, a diode is temperature dependent. Fig. 6.15a shows the block diagram of a rectifier using the Δ-ΣM approach.

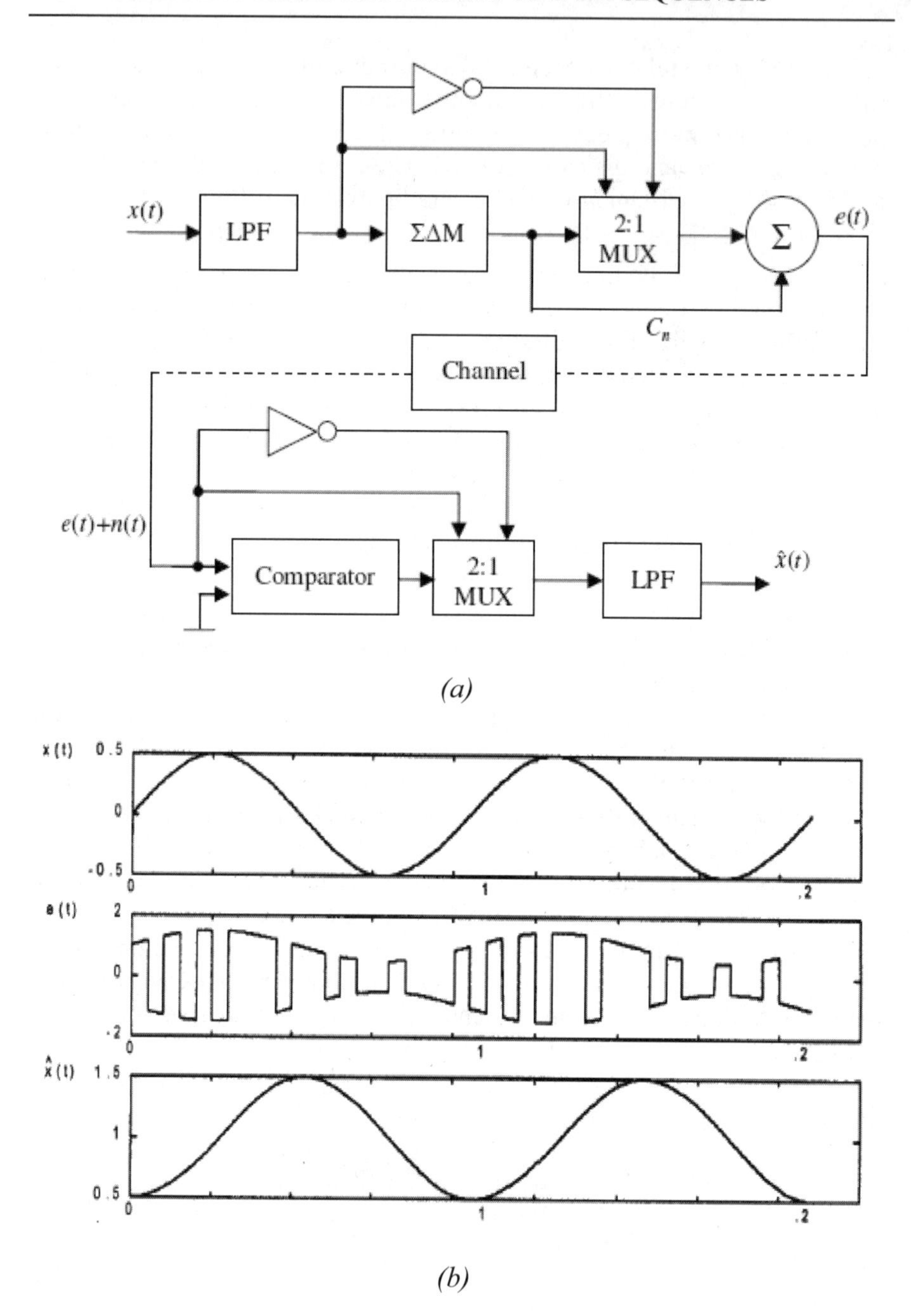
x(t)
LPF
ΣΔM
2:1 MUX
Σ
e(t)
C_n
Channel
e(t)+n(t)
Comparator
2:1 MUX
LPF
x̂(t)
(a)
x(t)
e(t)
x̂(t)
(b)

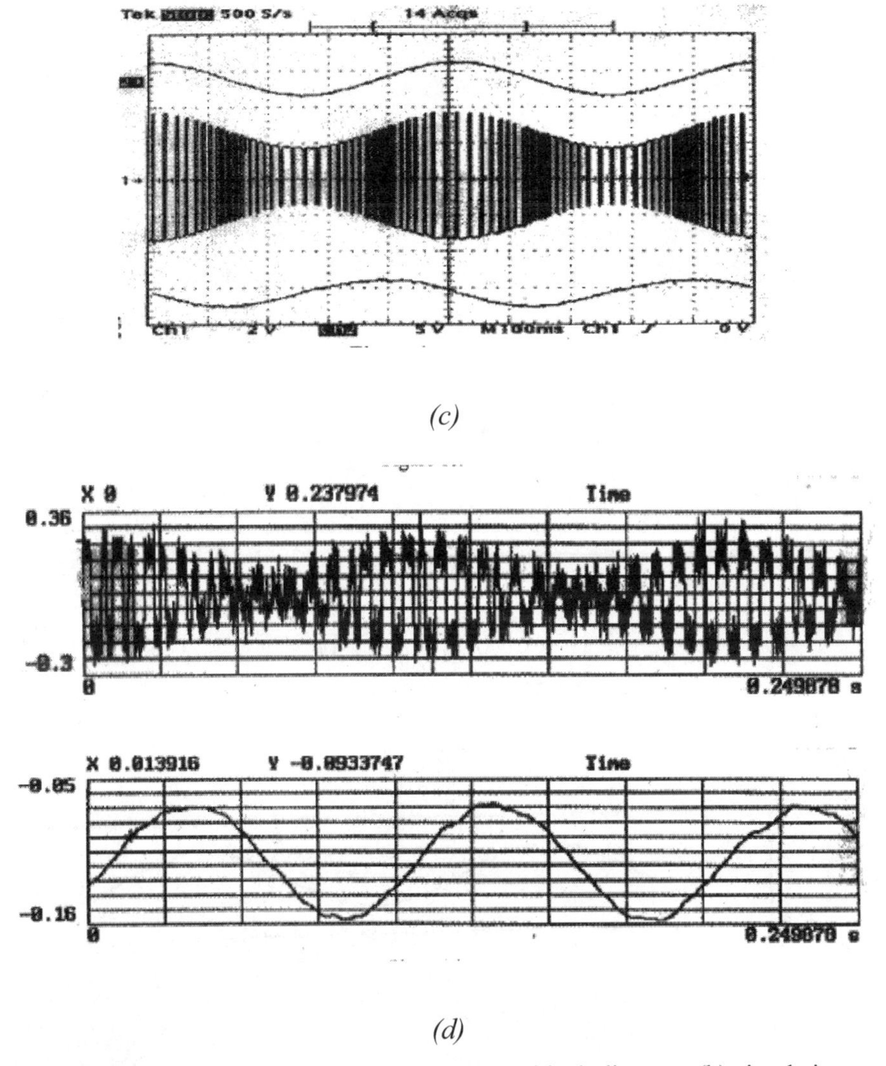


**Fig. 12.** Proposed asynchronous AM system, (a) block diagram, (b) simulation results, (c) experimental results, (d) experimental results with Gaussian noise added

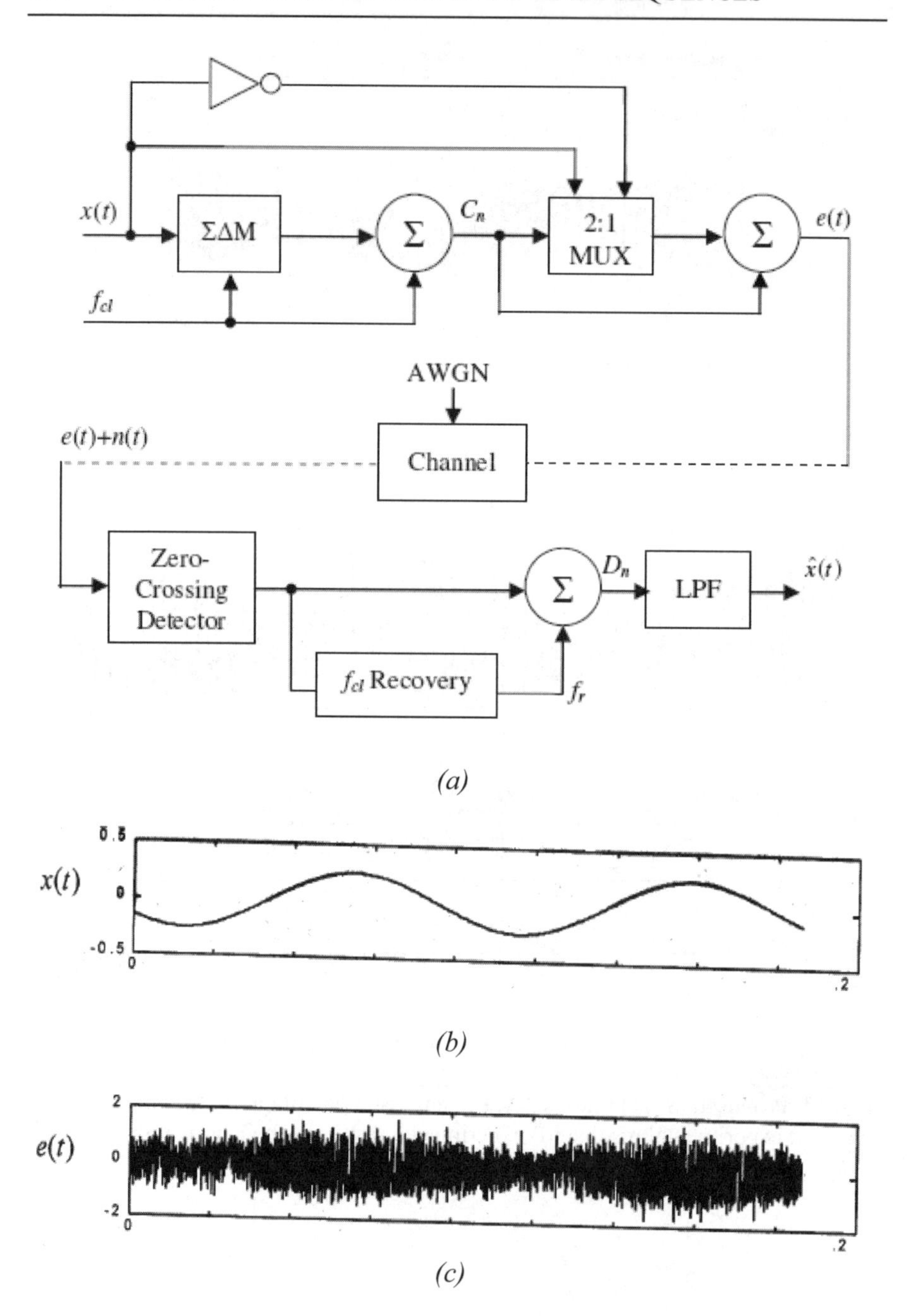
x(t)
ΣΔM
Σ
$C_n$
2:1
MUX
Σ
e(t)
$f_{cl}$
AWGN
e(t)+n(t)
Channel
Zero-
Crossing
Detector
Σ
$D_n$
LPF
$\hat{x}(t)$
$f_{cl}$ Recovery
$f_r$
(a)
x(t)
0.5
0
-0.5
0
.2
(b)
e(t)
2
0
-2
0
.2
(c)

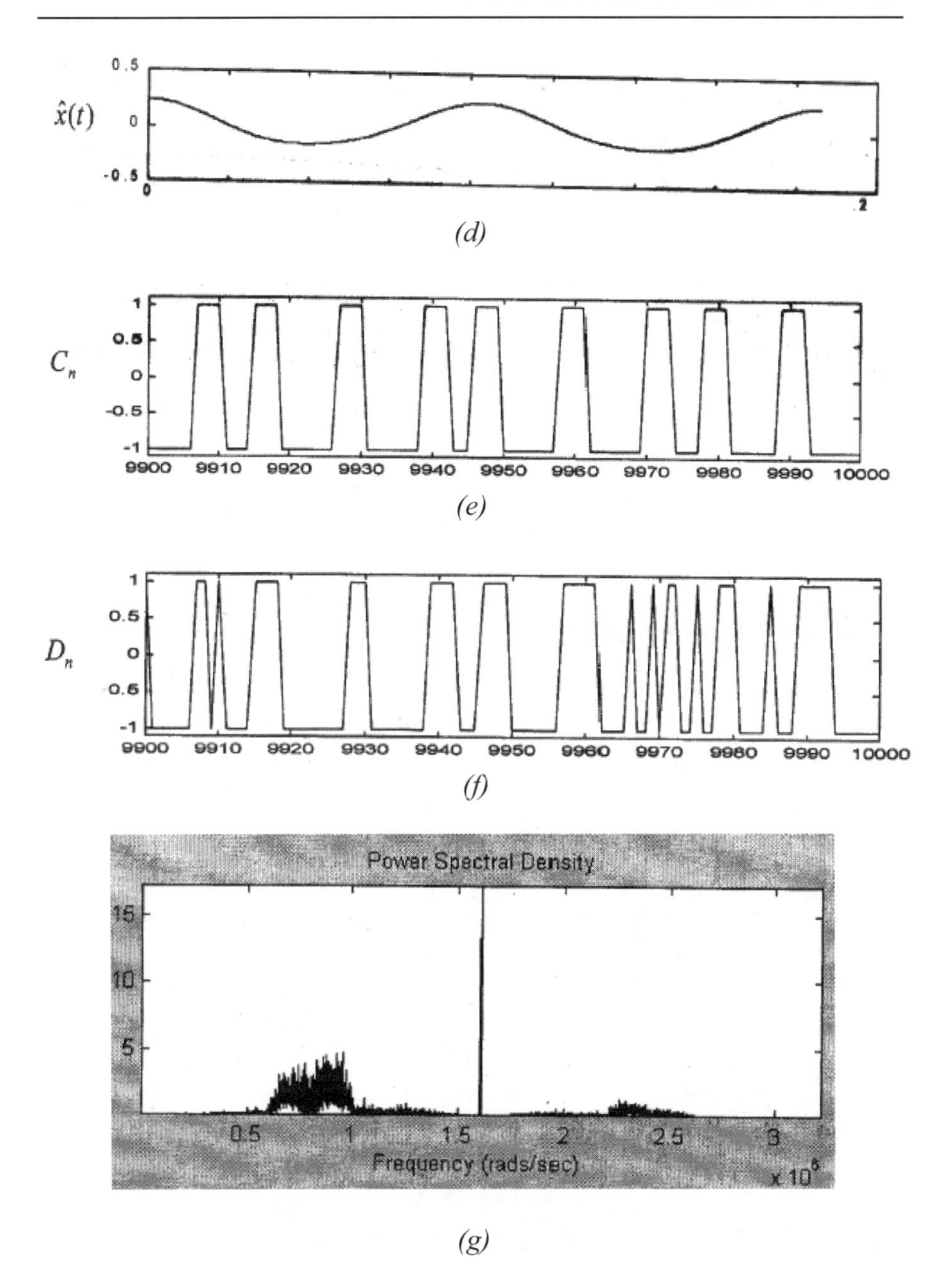


(g)

**Fig. 6.13.** Synchronous AM system and belonging waveforms

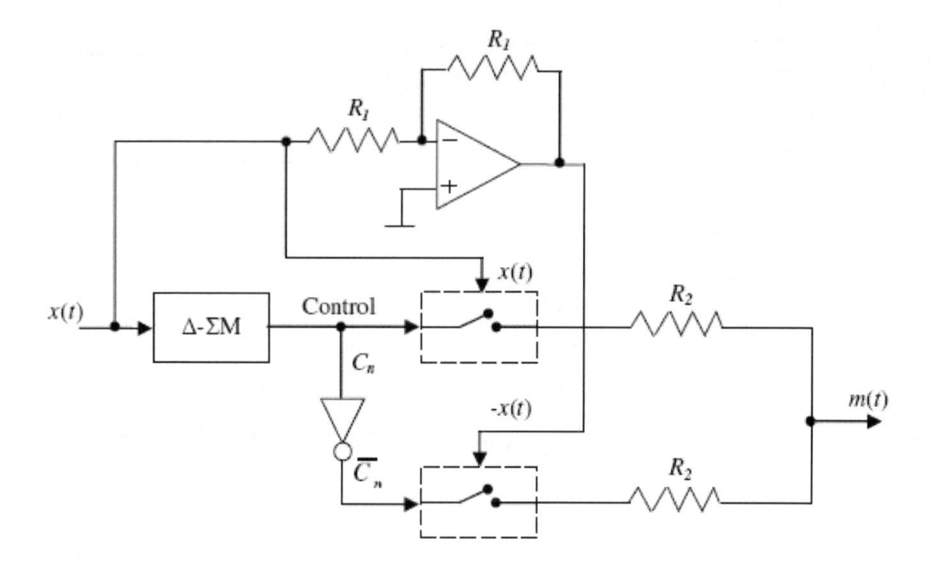


**Fig. 6.14.** Multiplexing multiplier

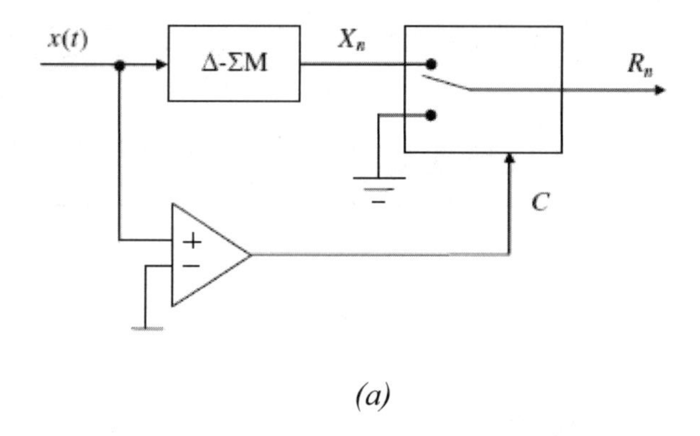


*(a)*

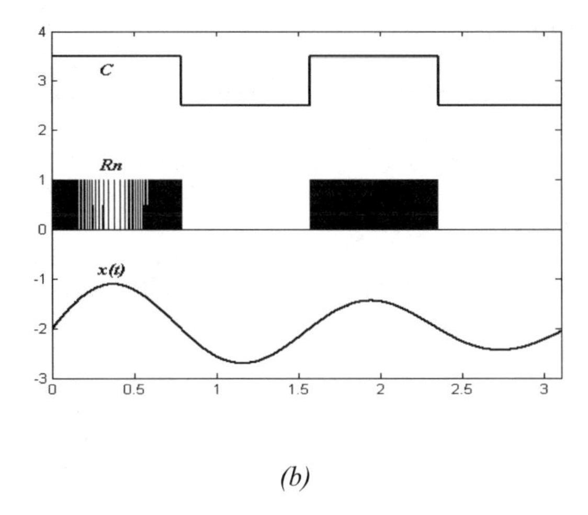


*(b)*

**Fig. 6.15.** (a) Block diagram of Δ-ΣM rectifier, (b) respective waveforms

## 6.5 CONCLUSION

In this chapter, the possibility of mixed analog/digital mode operations on delta-modulated pulse stream was introduced. Error analysis was done for different lengths of the averaging filter and different values of sampling frequency. A number of simulation examples were included to support our theoretical findings. An experimental AM system, based on delta modulation was also presented.

## REFERENCES

1. Zrilic, Dj., Circuits and Systems for Functional Processing of the Modulated Pulse Density Stream, US Patent # 6,285,306 BI.
2. Freedman, M., Zrilic, Dj., Nonlinear Arithmetic Operations on the Delta-Sigma Pulse Stream, Signal Processing – Elsevier Science Publishers B.V., Vol. 21, 1990, pp. 25-35.
3. Malboreti, F., "Non-Conventional Signal Processing by the Use of Sigma-Delta Modulation Techniques: A Tutorial Introduction," Proceedings of IEEE CAS, San Francisco, 1992, pp. 2645-2648.
4. Dias, V., F., "Sigma-Delta Signal Processing," Instituto Superior Tecnico, Seccao de Electrico, Av. Rovisco Pais, 1000 Lisboa Codex, Portugal.

# CHAPTER 7 DECODING OF FIRST-ORDER Δ-ΣM SEQUENCES

## 7.1 DECODING OF FIRST-ORDER Δ-ΣM SEQUENCES

### 7.1.1 Introduction

In general, a delta-sigma decoder is a linear or nonlinear low-pass filter whose role is to aggregate the "useful" signal spectral components and, at the same time, remove higher order spectral components of quantization noise. There are two basic methods of decoding delta-modulated signals, linear and nonlinear. The linear decoding method consists of classic analog or digital filtering, while the nonlinear method is based on principles of successive approximation and some initial conditions. In addition to the complexity and signal-to-quantization noise ratio of a decoder, important features of the decoder are synchronization and influence of errors on a decoded signal. In this chapter, we will analyze the performance of two linear decoders implemented as finite impulse response (FIR) filters. The first filter consists of uniform coefficients, and we will refer to it as "uniform FIR filter". The second filter proposed by Gray [1] is called "optimal FIR filter". In addition, the nonlinear decoder known as a ZOOMER will be analyzed as well [2].

### 7.1.2 Delta-Sigma Communication Model

For simulation purposes, the communication model of a delta-sigma system is presented in Fig. 7.1. Analog input signal $x(t)$ is sampled first with frequency $f_o$. Pulse amplitude samples are then fed into Δ-ΣM, whose binary output signal $Q(U_n) = +1$ if $U_n > 0$, and $Q(U_n) = -1$ if $U_n \leq 0$. Since channel error can occur during transmission, a reset signal is needed to assure the same initial conditions at coder and decoder. The decoder can be implemented as a linear or nonlinear filter. To have the correct decoding, three conditions have to be satisfied: (a) ideal bit synchronization

($\hat{f}_b = f_b$), (b) simultaneous reset of both coder and decoder, and (c) no channel errors, i.e. $\hat{Q}(U_0) = Q(U_0)$. We consider the case when only condition (a) is satisfied.

### 7.1.3 Delta-Sigma Decoder

In the case where channel errors don't occur and the ideal synchronization occurs, the output of the decoder can be written as

$$\hat{X} = F\{ Q(U_o), Q(U_1), \ldots, Q(U_{n-1})\}, \tag{7.1}$$

where $F\{\}$ is the function of the coder. In the case that the decoder is implemented as a FIR filter with coefficients $h(n)$, $0 \le n < N$ expression (7.1) can be written as

$$\hat{X} = \sum_{i=0}^{N-1} h(i) Q(U_{n-1-i}). \tag{7.2}$$

Here we analyze the following decoders

1. **Uniform filter** This is a filter with identical coefficients, i.e. $h(n) = 1/N$, for $0 \le n < N$. In this case, the decoder output $\hat{X}$ is the arithmetic mean value of the sequence $Q(U_o), \ldots, Q(U_{n-1})$.
2. **Optimal FIR filter** This filter is proposed by Gray [1]. The coefficients of this filter are given by

$$h_n = 6\frac{(n+1)(N-n)}{N(N+1)(N+2)}, \quad 0 \le n \le N-1. \tag{7.3}$$

   This filter has a symmetric impulse response, i.e. $h(N-i-n) = h(n)$
3. **"ZOOMER" decoder** This decoder belongs to the class of nonlinear filters and it was proposed by Hein and Zakhor [2]. Its design is based on the following assumption: Let the initial value of $U_o = 0$. Then, at an instant $n$, $U_n$ can be written as

$$U_n = \sum_{i=0}^{N-1} [x_i - Q(U_i)] = (\sum_{i=0}^{N-1} x_i) - S_n, \; n \ge 1, \tag{7.4}$$

   where $S_n$ is the sum of bits at coder output before instant $n$

$$S_n = \sum_{i=0}^{N-1} Q(U_i), \; n \ge 1. \tag{7.5}$$

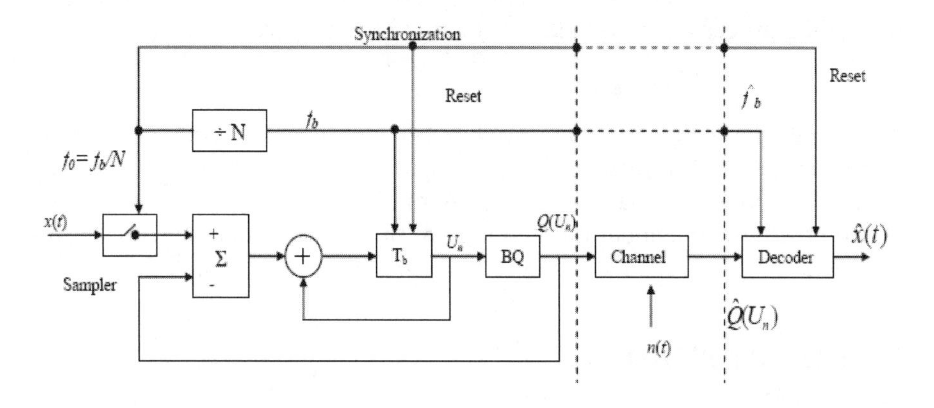


**Fig. 7.1.** Communication model of Δ-ΣM system [4]

Since the input signal to the coder is $x_i = x =$ constant for $i \geq 0$, the eqn. (7.4) can be written as

$$U_n = (\sum_{i=0}^{N-1} x_i) - S_n = nx - S_n \, , n \geq 1. \tag{7.6}$$

The sequence of $N$ bits is present at the input of the decoder $\{Q(U_n), 0 \leq n \leq N\text{-}1\}$, and the decoder knows its polarity only. Using eqn. (7.6), and knowing the polarity of $Q(U_n)$, it is possible to estimate the range of amplitude of the input signal $X$ at the moment of the generation of the digital word. Let $\overline{X}_n$ represent the mean value of the sequence $\{Q(U_i), 0 \leq i \leq N\}$ at the instant $n$, defined as

$$\overline{X}_n = \frac{1}{N} S_n . \tag{7.7}$$

Then, according to (7.6), every bit that arrives at the decoder defines one linear inequality, which gives an upper or lower boundary of the range of the input signal $X$. This means that at every instant $n$, it is possible to decide the upper or lower bound of the input signal in the following way

$$\text{If } Q(U_n) = +1, \text{ then } X > X_n, \tag{7.8}$$

$$\text{If } Q(U_n) = -1, \text{ then } X \leq X_n.$$

In this manner, for every code word, we can have $N$ linear inequalities. Solving this system of inequalities, it is possible to find the range of boundaries in which the input signal exists. Every signal from this range satisfies $N$ linear inequality equations. The same equation that describes the coding process is used in the decoding process as well, i.e.

$$U_n = U_{n-1} + X_{n-1} - Q(U_{n-1}),\ n \geq 1. \tag{7.9}$$

Thus, we can conclude that the input-output characteristic of such a nonlinear decoder is inversely proportional to the quantization characteristic of the coder, and decoding error is minimal. Minimal values of decoded error are dependent on the range of the transfer characteristic of the encoder.

### 7.1.4 Results of Analysis

To analyze the performance of decoders under consideration, the input signal of a certain level $X_k$ is encoded into the binary sequence $\{Q(U_o, \ldots, Q(U_{n-1})\}$. This sequence is then used to calculate the decoded value $\hat{X}_k$. A uniform distribution of input signal levels is assumed, i.e.

$$X_k = \frac{k}{2k+1},$$

where $|k| \leq K$, and $2k + 1$ is the total number of analyzed levels. Two parameters have been simulated and analyzed, absolute error

$$E_k = \left| X_k - \overline{X}_k \right| \tag{7.10}$$

and the mean value of the signal-to-quantization noise ratio,

$$SNR = 10\log_{10}\left(\frac{1}{2k+1}\sum_{k=-K}^{K} e^2{}_k\right) \quad dB. \tag{7.11}$$

1. **Ideal decoding**: Fig. 7.2 shows analysis results of $S/N$ as a function of encoded word $N$. As can be seen, ZOOMER algorithm performs better over the entire range of coded words for nearly six to eight decibels.
2. **Non-ideal decoding**: Fig. 7.3 represents the result of false synchronization when the input sample $X_k$ is periodically transmitted. It is evident that synchronization error has drastic influence on ZOOMER algorithm. For example, for $N = 128$, signal-to-noise ratio degrades for nearly fifty decibels. It can be concluded that uniform FIR filter performs best in the presence of imperfect synchronization.
3. **Influence of isolated errors**: The influence of isolated errors on the $SNR$ of a decoded signal is shown in Fig. 7.4, when $N = 128$ and $k = 1000$. The change of error position was performed in the range three to $N$ (the first two bits don't carry any information), then $SNR$ is calculated

according to eqn. (7.11) for the given input signal level $X_k$. The result of this analysis is shown in Fig. 7.4. We can see again the best performance of the uniform FIR filter independently of error position.

Performance of ZOOMER and optimal FIR filter depend on error position. Fig. 7.5 shows the results of simulation for SNR of decoded signal for different input levels. The SNR is calculated as the average of signal and noise power for all possible error positions. Again, we can see significant sensitivity of ZOOMER algorithm in the presence of only one error per coded word.

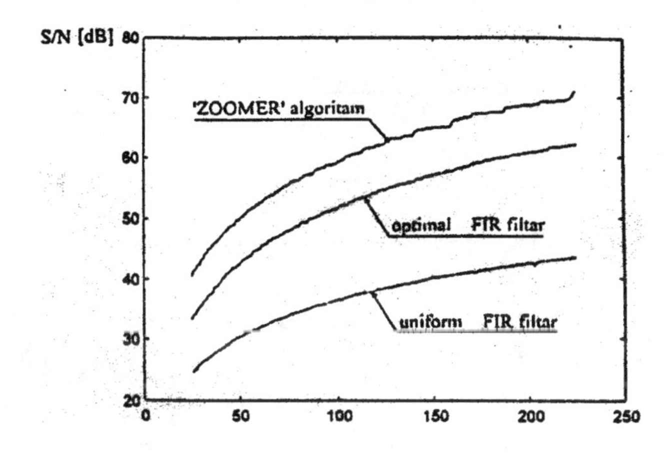


**Fig. 7.2.** Signal-to-noise ratio as a function of the length of coded word ($k$ = 1000) [4]

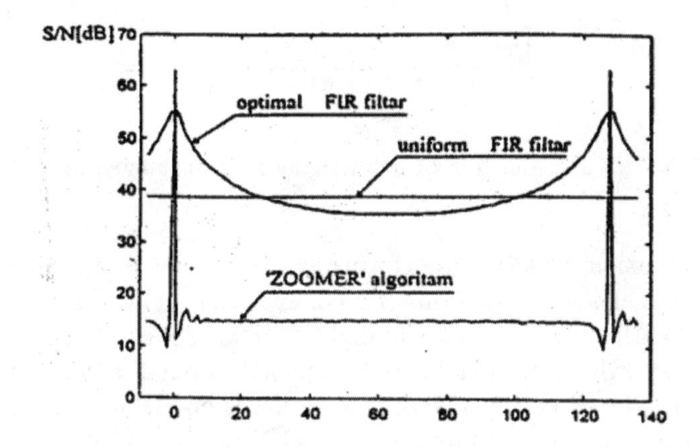


**Fig. 7.3.** Influence of synchronization error to SNR ($k$ = 1000, $N$ = 128) [4]

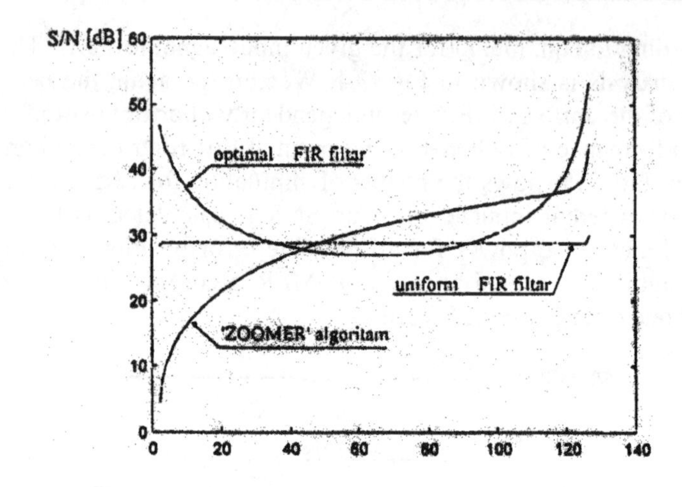


**Fig. 7.4.** Signal-to-noise ratio of decoding as a function of error position [4]

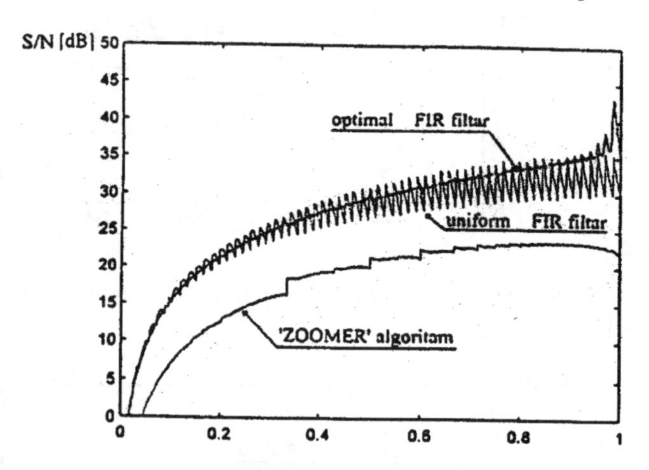


**Fig. 7.5.** *SNR* as a function of input signal level (averaged for all possible errors) [4]

In conclusion, observing performances of the three algorithms, we can see that under ideal conditions (no channel errors, ideal synchronization) the ZOOMER decoder has six to eight decibels better SNR in comparison to the Gray FIR filter, and 12 to 27 decibels better *SNR* than the uniform FIR filter. However, when synchronization is not achieved and when isolated channel errors are present, the ZOOMER decoder performs worst. Its *SNR* is worse by more than ten decibels. The ZOOMER decoder can be used in such communication systems where probability of error is negligible and synchronization is reliable. In the case of transmission through unreliable communication channels, use of FIR filters has an advantage.

## 7.2 SIMPLIFIED IMPLEMENTATION OF Δ-ΣM DECODERS

Here, a simplified structure of the ordinary delta-sigma decoder is described. The basic algorithm is derived and circuit diagrams for analog and digital implementation are proposed [3].

### 7.2.1 Basic Concept

The simplest delta-sigma decoder is an ordinary finite impulse response filter (FIR) with uniform coefficients. Let signals $X_n$ and $\hat{X}_n$ represent samples of the signal at the input of the encoder and the output of the decoder, respectively. When the delta-sigma decoder has the form of a FIR filter with uniform coefficients, as shown in Fig. 7.6. Its output is

$$\hat{X}_n = \frac{1}{N}\sum_{i=0}^{N-1} b_{n-i}\,, \tag{7.12}$$

where $Q(U_{n-1}) = b_n$ according to eqn. (7.9). Substituting eqn. (7.9) into (7.12), we can write $\hat{X}_n$ as the sum of two components

$$\hat{X}_n = \frac{1}{N}\sum_{i=0}^{N-1} b_{n-i} \quad + \quad \frac{(U_{n-N+1} - U_{n+1})}{N}\,. \tag{7.13}$$

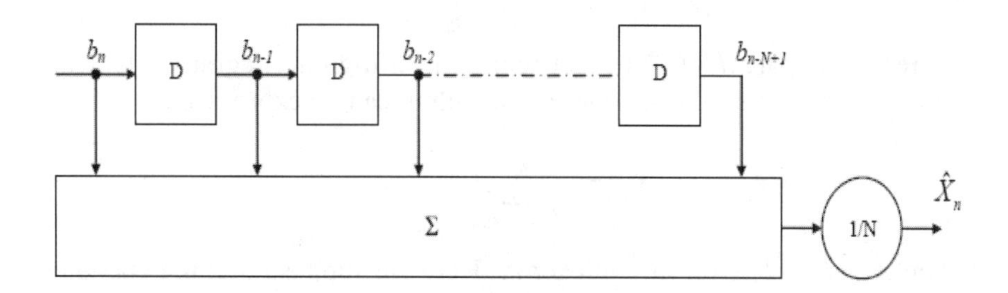


**Fig. 7.6.** FIR filter with uniform coefficients

The first component is the arithmetic mean of $N$ neighbor samples and the second component represents an "error". As was concluded in [1] and [2], if $|X_n| < 1$, then also $|U_n| < 1$. Thus, if the sampling frequency is much higher than the highest significant frequency components of the input signal and if the constant $N$ is large enough, then $\hat{X}_n$ will be a good approximation of $X_n$.

### 7.2.2 Implementation of the Delta-Sigma Decoder

A simple but naïve solution would be direct implementation of a FIR filter with uniform coefficients as shown in Fig. 7.6. A more sophisticated solution is to first write the transfer function of the uniform FIR filter. Using a D transform [1], this transfer function is given by

$$H(D) = \frac{1}{N}\sum_{i=0}^{N-1} D^i \,. \tag{7.14}$$

Expressing the sum on the right side in closed form we obtain

$$H(D) = \frac{1}{N}\frac{1-D^N}{1-D} = H_1(D)H_2(D)\,. \tag{7.15}$$

Based on this result, a delta-sigma decoder can be implemented as shown in Fig. 7.7. We see that the implementation of function

$$H_1(D) = 1 - D^N \tag{7.16}$$

consists of an $N$ bit long binary shift register and subtraction circuit. The digital recursive part with its transfer function is

$$H_2(D) = 1/N * (1 - D)^{-1}. \tag{7.17}$$

The recursive part, $H_2(D)$ in fact represents a digital integrator. It is not difficult to show that the samples at the output can be expressed as

$$\hat{X}_n = \frac{1}{N}\sum_{k=-\infty}^{n} d_k \tag{7.18}$$

where $d_k = b_k - b_{k-N}$ for any integer $k$. Here we suppose that the encoding and decoding processes start at instant $k = -\infty$.

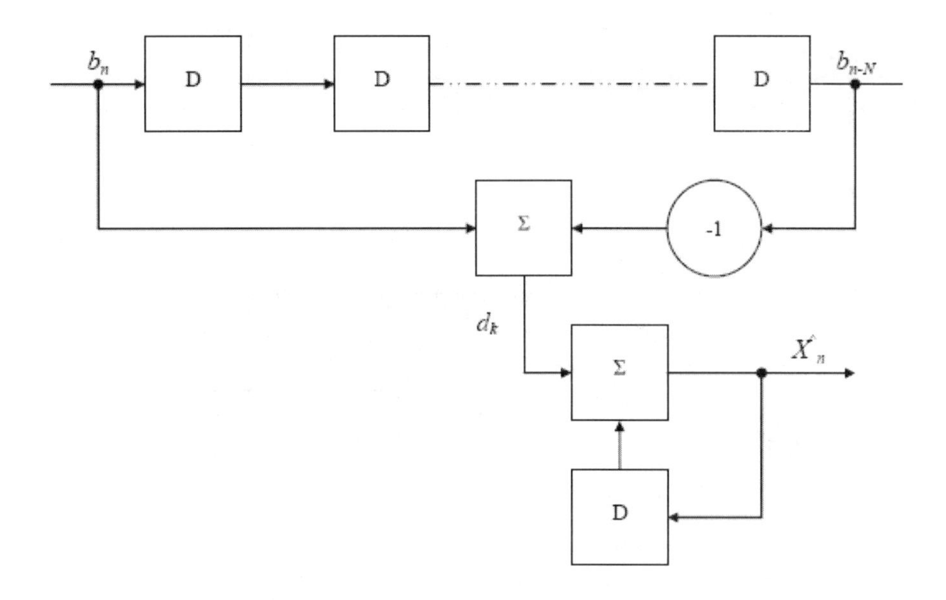


**Fig. 7.7.** Equivalent implementation of delta-sigma decoder

### 7.2.3 Proposed Implementation

Based on the concept explored in the previous section, we propose two solutions for implementation of the conventional delta-sigma decoder. Fig. 7.8 shows analog implementation of the decoder. The resistors $R_1$ and $R_2$ should have the same value. Other elements of the op amp, resistors $R_3$ and $R_4$ and capacitor C should be chosen to achieve a proper true constant and desired amplification of the whole integrator circuit. Fig. 7.9 represents a digital implementation of the delta-sigma decoder.

Considering binary values –1 and +1 as logic values “0” and “1” respectively, two AND gates and one XOR gate can be used to realize a count up when UP = “1”, count down when input DOWN = “1”, and stop when UP = DOWN = “0”. A binary counter consists of L flip-flops ($N \le L$). Outputs of this counter present pulse code modulation (PCM) words. These PCM words can be further used for additional digital signal processing with ordinary DSP hardware, or A/D converted into an analog signal.

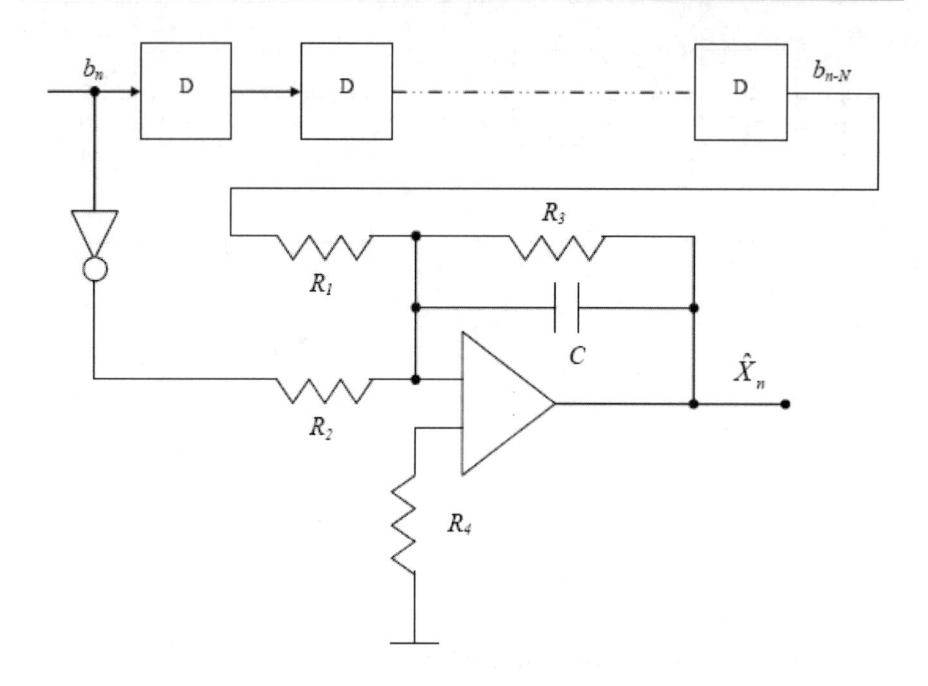


**Fig. 7.8.** Analog implementation of delta-sigma decoder [3]

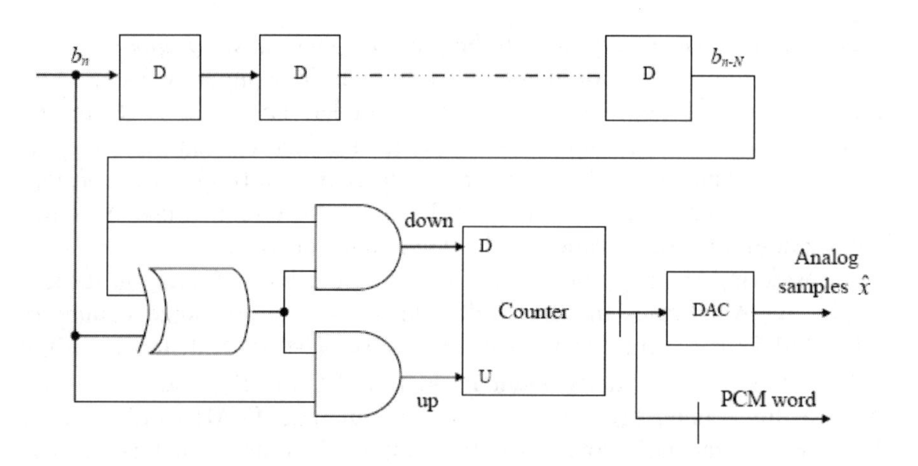


**Fig. 7.9.** Digital implementation of delta-sigma decoder [3]

## 7.3 CONCLUSION

In this chapter, we analyzed the performances of two delta-sigma decoders, FIR filter with uniform coefficients, and Gray's optimal FIR filter. signal-to-noise ratio was analyzed with and without channel errors. Results were compared and contrasted with performances of ZOOMER nonlinear decoder introduced in [2]. In addition, simplified implementation of a linear decoder was proposed.

## REFERENCES

1. Gray, R.M. "Spectral Analysis of Quantization Noise in a Single-Loop Sigma-Delta Modulator With DC Input," IEEE Trans. On Communications, 1989, COM-35, pp. 588-599.
2. Soren, H., Avideh, Z., "Sigma-delta Modulators Nonlinear Decoding Algorithm and Stability Analysis," Kluwer, Boston, 1993.
3. Zrilic, D.G., Yuan, B., Petrovic, G., "Simplified Realization of Delta-Sigma Decoder," *Electronics Letters*, 28 August,1997, Vol. 33, No. 18, pp. 1515 – 1516.
4. Stankovic, Z., Petrovic, G., Zrilic, D.G., "Comparative Analysis of Linear and Nonlinear Delta-Sigma Decoders," IV Telecommunication Forum, TELFOR '96, Beograd, November 26-28, 1996.

# CHAPTER 8 PCM – Δ-ΣM CONVERTERS

## 8.1 PCM - Δ-ΣM CONVERTERS

### 8.1.1 Introduction

The advantages of coding analog signals digitally are well known and widely discussed in the literature. Some well known coding schemas in practice are pulse code modulation (PCM), differential pulse-code modulation (DPCM), and delta modulation [1]. Waveforms coded in PCM involve sampling, quantization, and coding. The discrete amplitude levels of a pulse amplitude modulation (PAM) signal are represented by distinct binary words of length $n$. For example, with $n = 3$ one can represent 8 distinct levels. For decoding of a PCM signal, the binary words are mapped back into amplitude levels, and the amplitude-time pulse sequences are low-pass filtered with a filter of a certain cutoff frequency. For speech encoding, DPCM is frequently used as well. The Nyquist rate sampled speech exhibits a very significant correlation between successive samples. One consequence of this correlation is that the variance of the first difference $D_r(1) = X_r - X_{r-1}$ is smaller than the variance of the speech signal itself. As a result, it is advantageous to quantize $D_r(1)$ instead of $X$, and use an integrator to reconstruct $X$ from the quantized values of $D_r(1)$. Delta modulation exploits signal correlation in DPCM by over-sampling to increase the adjacent sample correlation. In fact, $\Delta M$ is a 1-bit version of DPCM and approximates an input time function by a series of linear segments of constant slope. Such a coder is therefore referred to as a linear delta modulator (LΔM). The drawback of the LΔM system is its sensitivity to the channel errors when LΔM pulses are sent over a transmission line. This problem was overcome by delta-sigma modulation (Δ-ΣM) [2], where the demodulator is an averager. There are a variety of code formats serving different terminals and transmission needs, and therefore, a need for code conversion. The Goodman [3] paper represents pioneering work in the field of achieving PCM conversion with a simple, non-adaptive, high bit rate LΔM. The delta modulation-to-PCM conversion method was also proposed by Kouvaras [7]. A simple and accurate digital converter was suggested that converts a delta modulated pulse density stream of an exponential delta modulator into a sequence of digital numbers. The proposed system employs an up-down counter with some logic based on a conventional full adder. Although there are different solutions for LΔM-to-PCM conversion, there is a need for PCM-to-Δ-ΣM conversion as well. This chapter describes two different methods of conversion of a PCM binary word in to a delta-sigma modulated pulse density stream.

### 8.1.2 Proposed Circuit Implementation

*A. Over-sampled PCM*

There are many different solutions for ΔΣM-to-PCM. One of them is proposed by Zrilic et al. in [4]. Our objective is to realize the PCM-to-Δ-ΣM marked with a bolded square in Fig. 8.1. The proposed logic block diagram of the PCM-to-Δ-ΣM converter is shown in Fig. 8.2. The function of this converter is described as follows. The output of the Δ-ΣM-to-PCM converter is fed into the block, for differential detection, and *N*:1 MUX. The differential logic block detects changes between two consecutive PCM words corresponding to the Δ-ΣM sampling rate. If change does not occur, the steering logic passes the output of the shift register to the register's input; otherwise, the output of the multiplexer is fed to the register. In the latter case, the address logic should pass a changed bit with the highest weight inside the PCM word. This means that if changes are present, the new Δ-ΣM bit describes the direction of change. The length of the shift register has to be the same as the register length in the Δ-ΣM-to-PCM converter proposed in [4].

To verify the validity of the proposed schema, the bread-boarding of the system from Fig. 8.1 was completed and Δ-ΣM ADMOD79 JQ was employed as an A/D converter. An input sinusoidal signal of 10 Hz, a sampling frequency of 20 kHz, a third-order low-pass filter, and a cut-off frequency of 20 Hz were used. Fig. 8.3 shows good agreement between the input signal $x(t)$ and the reconstructed signal $\hat{x}(t)$.

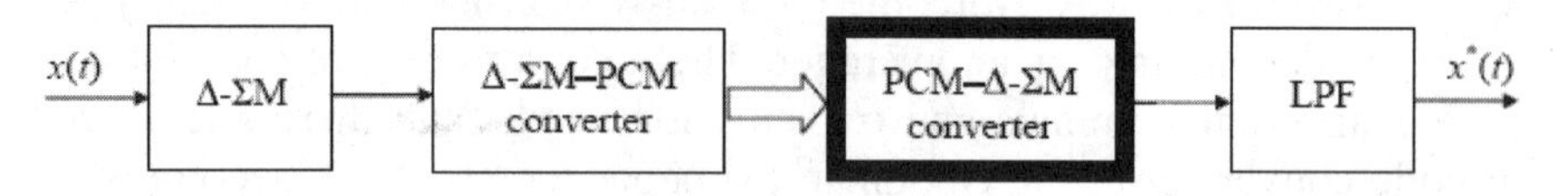


**Fig. 8.1.** Block diagram for the back to back conversion technique

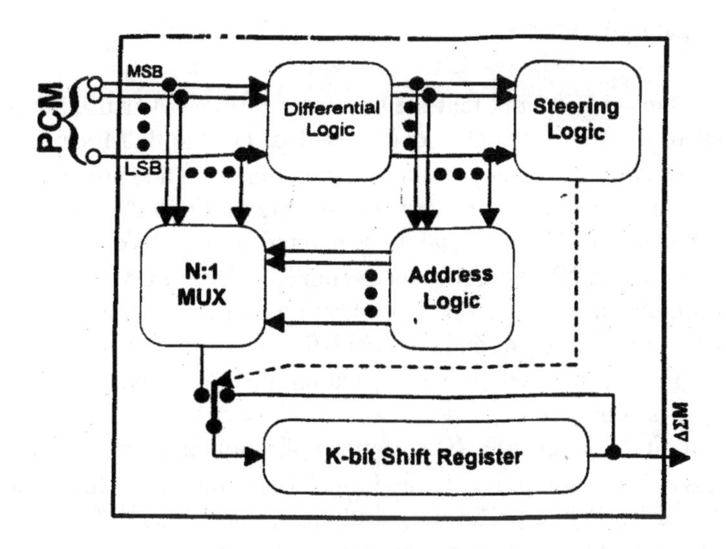


**Fig. 8.2.** Proposed block diagram of PCM-to-Δ-ΣM converter [8]

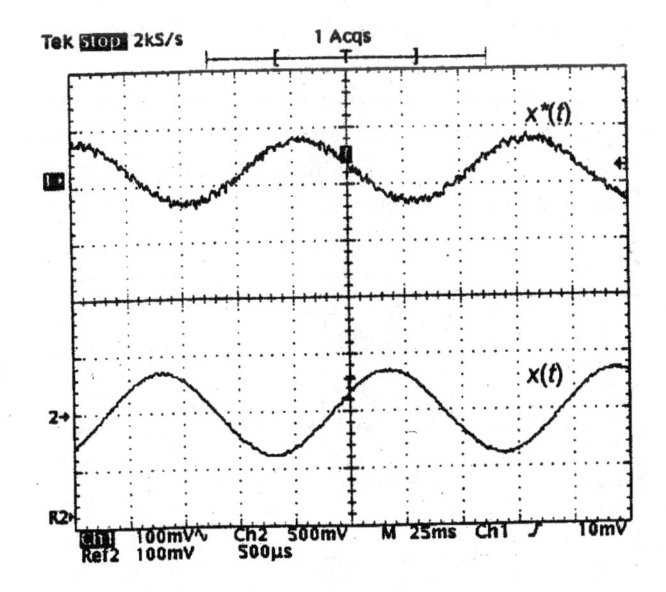


**Fig. 8.3.** Signal reconstructed after the double conversion (Δ-ΣM-to-PCM and PCM-to-Δ-ΣM) according to fig. 1 [8]

*B. Ordinary PCM*

It is important to point out that the converter in Fig. 8.1 is used in applications where the PCM signal is over-sampled, i.e. the PCM sample rate is equal to the clock frequency of the Δ-Σ modulator. The application of this type of converter therefore, can be somehow limited. We propose a method where the PCM signal is generated at the Nyquist (or a little higher) rate and Δ-ΣM is highly over-sampled. Both PCM and Δ-ΣM signals are digital in nature, so it is proper to assume that conversion can be performed in the same domain without the need to take an excursion to analog signal reconstruction. The digital circuit for regular PCM-to-Δ-ΣM conversion is shown in fig. 8.4.

All circuits' blocks are synchronized through a common clock that represents a Δ-ΣM sampling rate and it is independent from the sample rate of PCM words. While performing arithmetic operations in the binary system a complement of two is needed for subtraction. A necessary complement is found in the first block of the diagram in Fig. 8.4. Together with the full adder, it performs subtraction of the input PCM word from the reference level on the ROM output. The next two blocks are introduced for iterative summation of consecutive differences. Depending on the sum sign, a 1 or 0 is sent to Δ-ΣM stream, and the appropriate level is set on the ROM output. This information enforces negative feedback that keeps the registers content around zero, oscillating between positive and negative numbers. Conceptually, this is very alike to the analog version of the Δ-Σ modulator done with binary arithmetic.

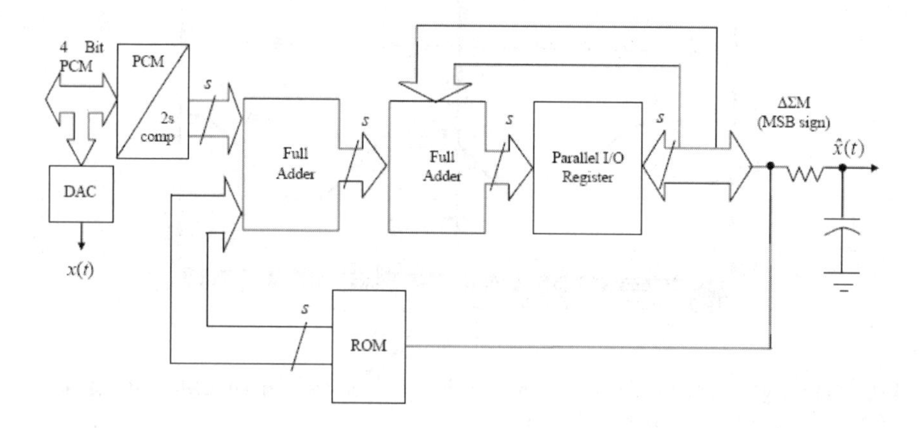


**Fig. 8.4.** Block diagram of ordinary PCM-to-ΔΣM conversion [8]

Fig. 8.5 shows the experimental test result when a 4-bit linear PCM word is converted into a serial Δ-ΣM pulse stream. The PCM word is generated by a binary ripple counter being clocked with pulses of frequency 500 Hz. The analog equivalent of this word is used for control purposes and shown as $x(t)$. For a Δ-ΣM sampling rate of 20 kHz, $\hat{x}(t)$ shows an analog signal reconstructed from the produced Δ-ΣM stream. This signal is obtained by averaging the Δ-ΣM signal with a first-order integrator. The quality of the reconstructed signal depends on the sampling frequency of the PCM-to-Δ-ΣM circuitry and the order of the averaging filter. From Fig. 8.5, it can be seen that signals $x(t)$ and $\hat{x}(t)$ are almost identical, which proves the possibility of a direct conversion.

## 8.2 DIGITAL-TO-ANALOG CONVERTER BASED ON ΔM

### 8.2.1 Introduction

The need for faster digital to analog converters is much greater than in the past. There exists today a wide range of applications for DACs: instrumentation, CAD systems, image processing, direct digital waveform synthesis, etc. The usual deciding factors in choosing a DAC are resolution and speed. The faster the DAC, the higher the resolution that can be attained. Most of today's digital to analog converters include additional digital support functions. However, the performance of an analog signal can be degraded by additional digital circuitry, and in mixed digital-analog systems there are inevitable compromises. As a step toward a partial solution to this problem, digital circuits are used to implement a new type of DAC.

A conventional digital to analog conversion involves analog voltage division (by two) and summation. For this process, well matched passive components are used. The processing costs for linear nickel-chrome resistors or double poly-silicon capacitors are relatively high. Alternatively, the low cost of first order linear delta-modulator (LΔM) encoders make them attractive for signal processing applications. With a low cost digital network for direct arithmetic operations on a LΔM pulse stream, it is possible to build new DAC structures. A voltage-mode DAC technique is implemented with digital circuits, which eliminates the need for passive components in performing voltage scaling.

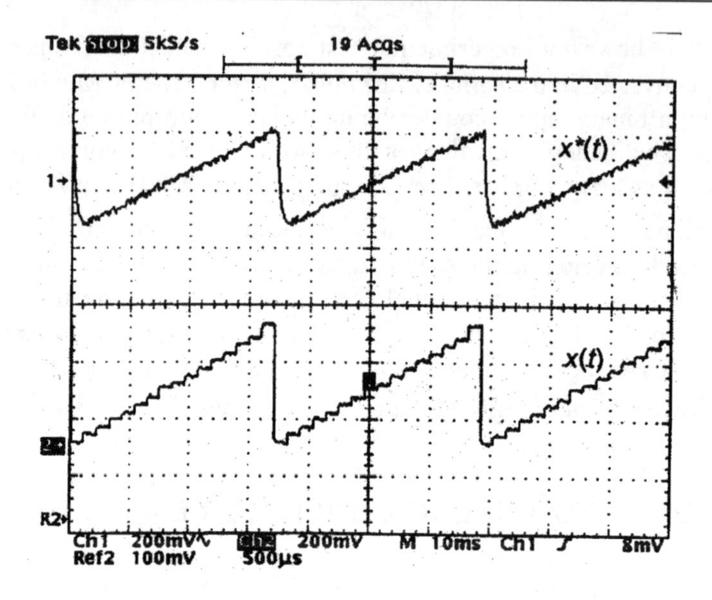


**Fig. 8.5.** Reconstructed analog signal after the PCM - Δ-ΣM conversion, $\hat{x}(t)$, and analog representation of the original PCM word, $x(t)$ [8]

### 8.2.2 A New DAC

A commonly used DAC has a voltage summing or R-2R ladder network, as shown in Fig. 8.6. The output voltage for any particular binary input is

$$E_0 = E_{ref} \sum_{j=1}^{n} \frac{a_j}{2^j}, \tag{8.1}$$

where $a_j$ is equal to either one or zero. To ensure high speed operation, fast voltage adders are required. The gain of the voltage adder is the dominant parameter that affects the differential and integral linearity of the DAC. Modest accuracy requires the use of a voltage adder with reasonable, well-controlled gain. Instead of dealing with operational amplifiers and a number of well-matched resistors, a more promising approach is to use the binary delta-adder.

#### The Binary Sequence Signal Processing Element

N. Kouvaras [6] showed that a binary delta-adder may be realized simply as shown in Fig. 8.7. The discrete sequences $X_n$ and $Y_n$ are synchronous sigma-delta modulated sequences. The basic building elements of the two-input delta-adder are the conventional binary full-adder (FA), and the D flip-flop, interchanging the role of SUM and CARRY of the ordinary FA, as seen in Fig. 8.7.

Assuming $\{X_n\}$ and $\{Y_n\}$ to be binary delta-sigma sequences, then the sequences $\{S_n\}$ and $\{C_n\}$ are also binary delta-sigma sequences. Using well known FA equations, it can be shown that

$$S_n = \frac{1}{2}[X_n + Y_n - (C_n - C_{n-1})] \qquad (8.2)$$

$$C_n = X_n Y_n C_{n-1},$$

where $X_n$, $Y_n$, $C_{n-1}$ are {-1, +1} and $n = \ldots -1, 0, +1, \ldots$ . Summing the left and the right sides of eqn. 8.2 and multiplying by the delta-step size $\delta$, it can be shown that

$$\delta\sum_{j=k}^{n-1} s_j = \frac{1}{2}\delta\sum_{j=k}^{n-1}(X_j + Y_j) + \frac{1}{2}\delta\sum_{j=k}^{n-1}(C_{j-1} - C_j),\ k < n - 1, \qquad (8.3)$$

where the theory is valid as $k$ approaches infinity. Notice that

$$\hat{S}(n\Delta t) = \delta\sum_{j=k}^{n-1} s_j$$

represents the delta demodulated signal, where $\Delta t$ is the sampling interval. Introducing the substitution

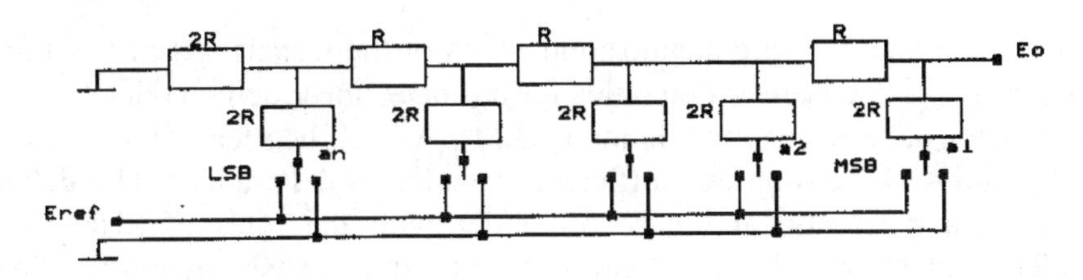


**Fig. 8.6.** Voltage-summing DAC

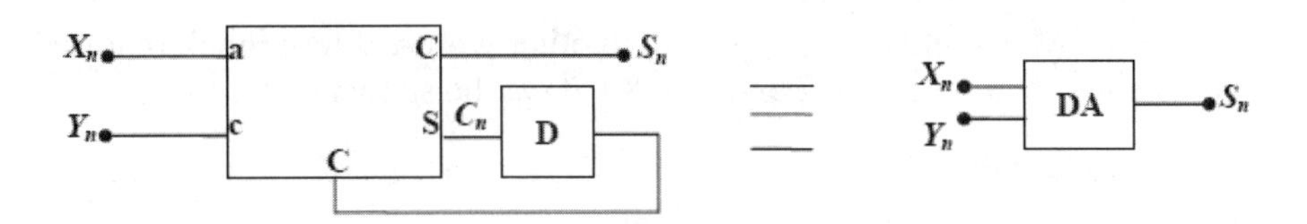


**Fig. 8.7.** Circuit configuration of delta adder

$$\Theta(n\Delta t)=\frac{1}{2}\delta\sum_{j=k}^{n-1}(C_{j-1}-C_j)=\frac{1}{2}\delta(C_k-C_{n-1}), \tag{8.4}$$

eqn. 8.3 can be written as

$$\hat{S}(n\Delta t)=\frac{1}{2}[\hat{x}(n\Delta t)+\hat{y}(n\Delta t)]+\Theta(n\Delta t), \tag{8.5}$$

or for a continuous time waveform, the delta demodulated signal is

$$s(t)=\frac{1}{2}[x(t)+y(t)]+\Theta(t), \tag{8.6}$$

where, $n\Delta t \le t \le (n+1)\Delta t$. It can be seen that the demodulated signal $s(t)$ is equal to one-half of the sum of the demodulated signals $x(t)$ and $y(t)$ plus some error $\Theta(t)$. Since $C_j$ can have a value of +1 or –1, the absolute value of the error satisfies the inequality,

$$|\Theta(t)|\le\delta. \tag{8.7}$$

If we take into consideration the quantization errors of the signals $x(t)$ and $y(t)$ by calculating their half sum, one obtains the error,

$$|E_s(t)|=|\Theta(t)|+\frac{1}{2}|E_x(t)+E_y(t)| \tag{8.8}$$

where $E_x(t)$ and $E_y(t)$ are quantization errors of the signals $x(t)$ and $y(t)$ respectively. This consideration holds for ordinary linear delta modulation as well as for delta-sigma modulation (Δ-ΣM). As will be seen, a binary full-adder and a D flip-flop are sufficient to realize a delta-adder. The delta-adder is a conventional binary full-adder with the roles of SUM and CARRY interchanged. This circuit can be used as a basic circuit-building block to perform voltage division at the digital circuit level. With a low cost digital network for direct arithmetic operations on a LΔM pulse stream, it is possible to implement a new structure for a digital to analog converter, as shown in Fig. 8.8.

If the Δ-ΣM is highly over-sampled, in other words, if the signal to noise ratio is high, then by eqn. 8.2 and Fig. 8.8 it can be shown that

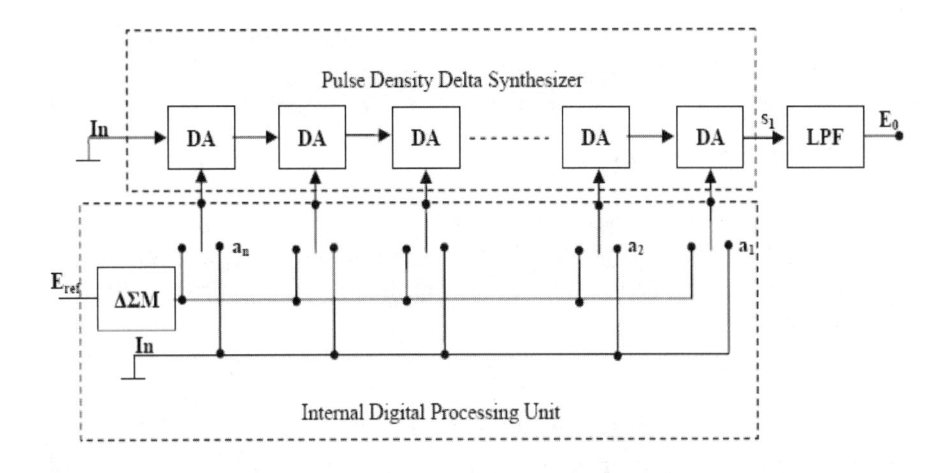


**Fig. 8.8.** Block diagram of proposed Δ-ΣM DAC [9,10]

$$S_n = \frac{1}{2}[a_n + I_n] \tag{8.9}$$

$$S_{n-1} = \frac{1}{2}[S_n + a_{n-1}] = \frac{1}{4}a_n + \frac{1}{2}a_{n-1} + \frac{I_n}{4}$$

$$\vdots$$

$$S_1 = \frac{1}{2^n}I_n + \frac{a_n}{2^n} + \frac{a_{n-1}}{2^{n-1}} + \ldots + \frac{a_2}{2^2} + \frac{a_1}{2^1},$$

where $I_n$ is the idle sequence, defined by $I_n =$ ...-1,+1,-1,+1, ... After delta demodulation (low pass filtering) of the sequences $\{S_n\}$, the output voltage $E_0$ is given as

$$E_0 = E_{ref}[\frac{a_1}{2^1} + \frac{a_2}{2^2} + \ldots + \frac{a_n}{2^n}] \tag{8.10}$$

or

$$E_0 = E_{ref}\sum_{j=1}^{n}\frac{a_j}{2^j} \tag{8.11}$$

which is the same as eqn. 8.1. This is the case when $E_{ref}$ is delta-modulated, and the digital input word is any kind of pulse-code modulation. In fact, the digital word to be converted plays a role of control input to the pulse density synthesizer, Fig. 8.8.

### 8.2.3 Simulation Results

A first order Δ-ΣM is simulated to modulate $E_{ref}$. The over-sampling ratio was 1000 ($R = f_s/2f_c$). To get the demodulated output of $E_0$, as seen in Fig. 8.8, averaging over 512 bits was used. Fig. 8.9 shows the process of adding two signals.

The signal $s(t)$ is the demodulated sum of two Δ-ΣM signals, $X_n$ and $Y_n$, corresponding to analog inputs $x(t)$ and $y(t)$. In Fig.8.9, the error signal of addition is $E(t) = s(t) - \hat{s}(t)$, which is present when the averager length is 512 bits. In Fig. 8.9, the error signal represents the case when the length of the averager is 256 bits. It is easy to see that the length of the averaging filter is of crucial importance for the reduction of error.

In Fig. 8.10, the output signal $E_0$ of the 3-bit Δ-ΣM-DAC from Fig. 8.8 is shown. Simulation results indicate that length of the LP Filter is proportional to $2^n$, where $n$ is the length of the binary code word to be converted.

An interesting case is when the Δ-ΣM of Fig. 8.8 is replaced by a clock-controlled switch. The frequency of switching between $-E_{ref}$ and $+E_{ref}$ is $f_s$ as well. Fig. 8.11 shows the output $E_0$ of a 3-bit DAC when the sampling frequency of the switch is $f_s$ = 1000 Hz. It can be seen that this result is identical to the case of Fig. 8.10 when Δ-ΣM is used.

The Δ-ΣM-DAC can be used as a multiplying DAC, and instead of $E_{ref}$, any type of signal can be used (This signal has to be highly over-sampled as well). Fig. 8.12 shows the case when a sinusoidal input signal is Δ-ΣM encoded and multiplied with a constant of $n = 3$ bits.

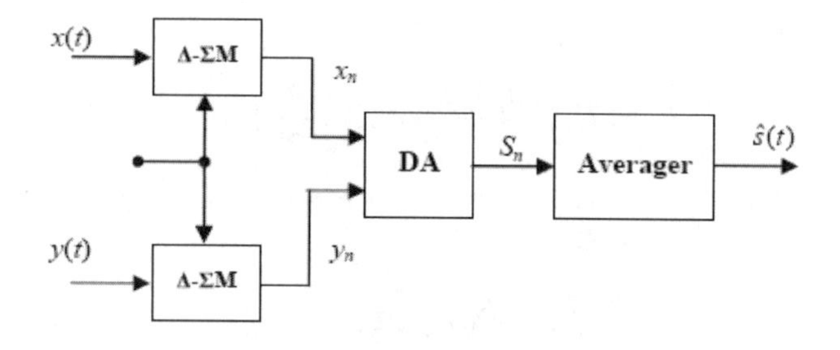

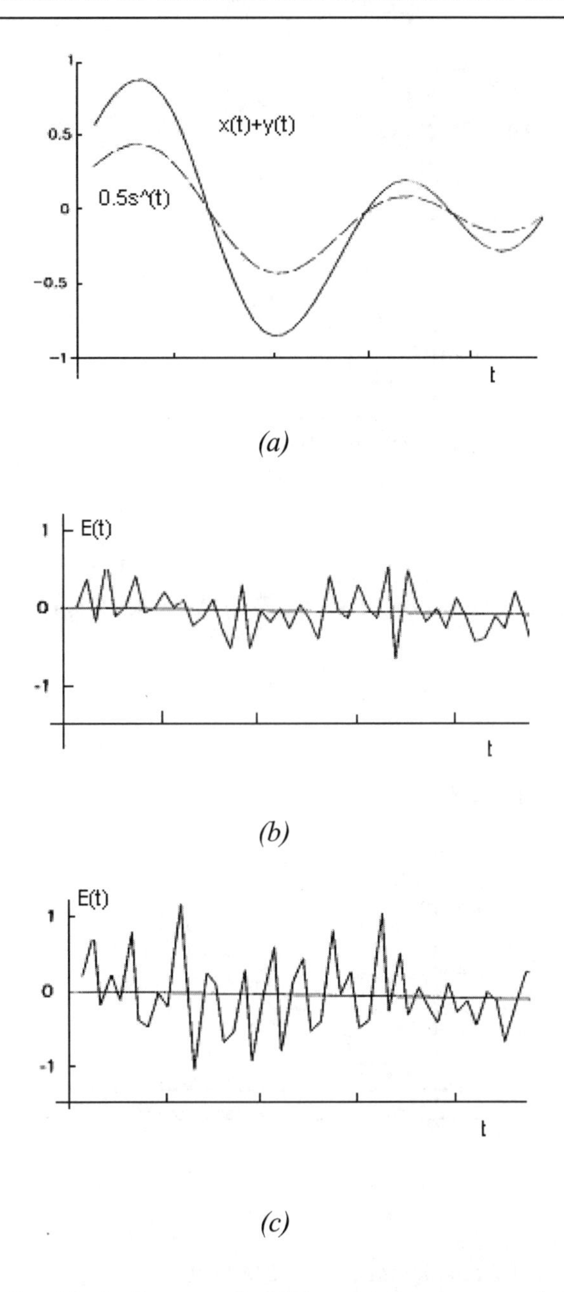


**Fig. 8.9.** Simulation block diagram of addition, (a) Input sum and demodulated output sum $0.5\,\hat{s}(t)$, (b) Magnitude of error when the length of averager is 512 bits, (c) Magnitude of error when the length of averager is 256 bits [9]

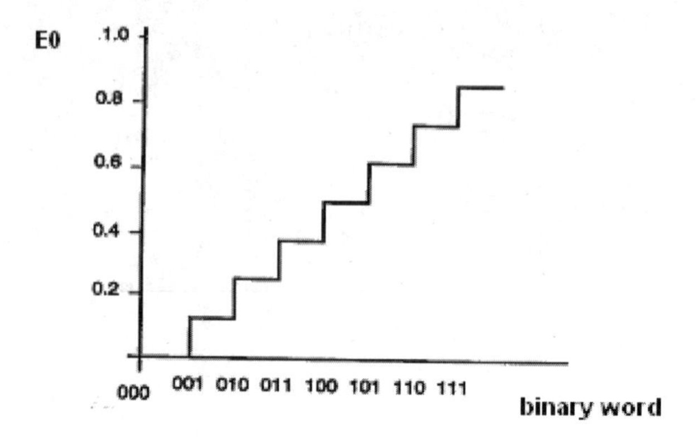


**Fig. 8.10.** Transfer function of 3-bit ΔΣM DAC when $E_{ref}$ is Δ-ΣM pulse-stream [9]

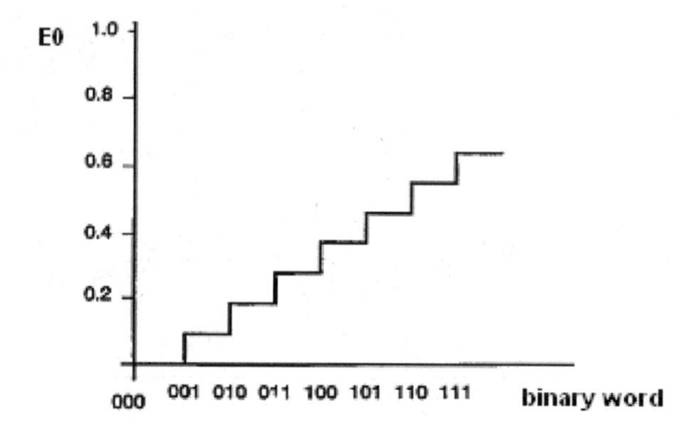


**Fig. 8.11.** Transfer function of 3-bit Δ-ΣM DAC when $E_{ref}$ is switched between –$E_{ref}$ and +$E_{ref}$ [9]

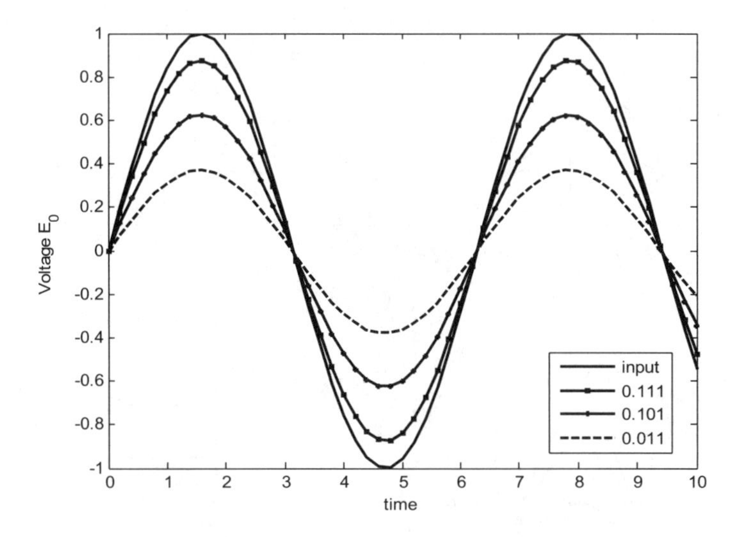


**Fig. 8.12.** Output of multiplying ΔΣM-DAC when $E_{ref}$ is a sinusoidal signal

Another interesting use of the Δ-ΣM-DAC is in the application of multiplying two digital words. Fig. 8.13 shows a proposed DAC configuration.

It shows the case when the input signal is $e^{-t/10}\sin(t)$. This signal is sampled and delivered to the digital input of Δ-ΣM-DAC1 as 8 bit PCM word. $E_{01}$ is the pulse density modulated (PDM) signal and is the reference signal for Δ-ΣM-DAC2. A digital word of 3 bits is delivered to the input of Δ-ΣM-DAC2. The results of multiplication of this example are presented in Fig. 8.13. The averager length also was 512 bits. In all diagrams, the $Y$ and $X$ axes represent normalized voltage and time, respectively.

We can conclude that to reduce the amount of power and area needed in an integrated circuit version of a DAC, a one bit interpolative Δ-ΣM DAC can be used. The basic building block of this DAC is an ordinary binary full-adder with interchanged roles of SUM and CARRY. The simulation results indicate that conversion relies on the interchangeability of the converter resolution and sampling rate. The lab experiments show that realization is possible in any technology (TTL, CMOS, etc.) The remaining LP filter is essentially the same smoothing filter needed for any multi-bit DAC. It is worth mentioning here that Δ-ΣM systems are inherently highly over-sampled, a fact which relaxes the requirements for low-pass filter design.

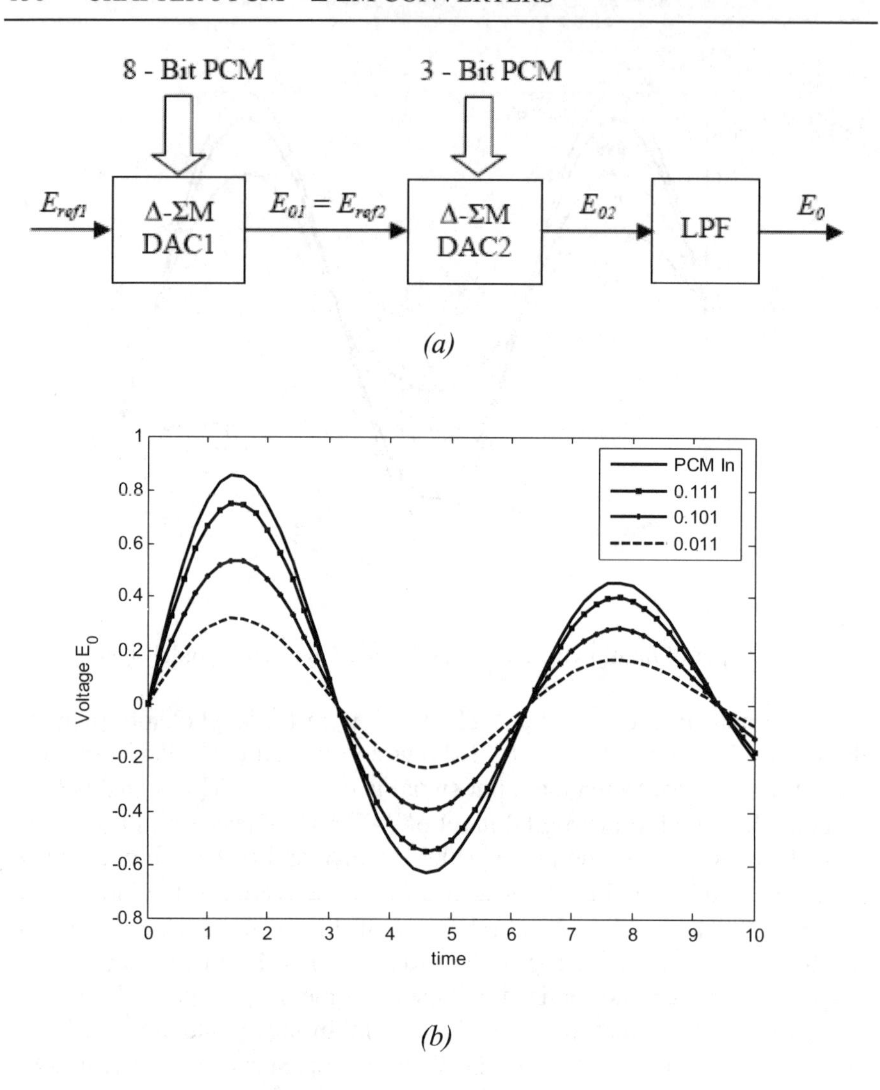


**Fig. 8.13.** (a) Block diagram of multiplication of two digital numbers using Δ-Σ Multiplying DAC, (b) belonging waveforms [10]

## 8.3 CONCLUSION

Our conclusion is that direct digital conversion among code formats offers better performance, flexibility, and economy. The conversion between PCM and Δ-ΣM formats can be implemented with standard logic circuits. In addition, we have introduced a novel type of DAC, based on arithmetic operations on Δ-ΣM pulse density stream.

## REFERENCES


1. N. Jayant, Editor, *Waveform Quanitization and Coding*, IEEE Press, 1976.
2. H. Inose, Y. Yasuda, J. Murkami, "A Telemetring System by Code Modulation -ΔΣ Modulation", *IRE Trans. On Space Electronics*, Telemetry, vol. 8, pp.204-209, Sept. 1962.
3. D.J. Goodman, "The Application of Delta Modulation to Analog-to-PCM Encoding", *Bell Systems Technical Journal*, vol.48,pp.31-343, Feb.1969.
4. Dj. Zrilic, G. Petrovic and B. Yuan, "Simplified Realization of Delta-Sigma Decoder" *Electronics Letters*, 1997, vol.33, no.18, pp.1515-1516.
5. J. Child, "High-Speed DACs Target Waveform Synthesis and Video", *Computer Design*, Feb.1, 1991, pp. 93-97.
6. N. Kouvaras, "Operations on delta-modulated signals and their applications in the realization of digital filters," *The Radio and Electronics Engineer*, vol. 48, no.9, Sept. 1978, pp.431-438.
7. N. Kouvaras, "Delta-Modulation/P.C.M. Converter", Electronics Letters, 28th September 1978, Vol. 14, No. 20, pp. 660-662.
8. Dj. Zrilic, M. Narandzic, "PCM-Δ-ΣM Converter", 42nd Midwest Symposium on Circuit and Systems, Las Cruces, NM, August 8-11, 1999, pp. 1036-1038.
9. Dj. Zrilic, "A New Digital-to-Analog Converter based on Delta Modulation", Midwest Symposium on Circuits and Systems, August 3-5, 1994, Lafayette, Louisiana, pp. 1191-1195.
10. Dj. Zrilic, "Methods and Apparatus for Mixed Analog and Digital Processing of Delta Modulated Pulse Stream Including Digital-to-Analog Conversion of Digital Input Signal", US Patent #5,349,353.

# CHAPTER 9 STOCHASTIC PROCESSING USING Δ-ΣM

## 9.1 INTRODUCTION

The basic idea of stochastic processing is to use probabilities as information carriers. The information is carried by the probability of occurrence of a "HIGH" logic level. Each logic level is statistically independent and it has the form of a Bernoulli sequence. It was shown in [11] that arithmetic operations of inversion, multiplication, addition and integration of discrete variables are possible if statistical independence of discrete events is assured. To assure statistical independence, the design of a suitable stochastic analog-to-digital converter is needed whose output pulses are uniformly distributed. The various approaches of random number generation are described in [1] and elsewhere. It is desirable that the output of a stochastic analog-to-digital converter has the form of a non-stationary Bernoulli sequence. In addition to non-stationary Bernoulli generation, [11] gives examples of possible arithmetic circuit realization. An example of digital to stochastic conversion is shown in Fig. 9.1.

We can estimate the probability $p$ by considering the frequency of occurrence of a logic "high" event in time intervals $N$, where $N = 2^n$, the length of linear feedback shift register (LFSR). Then

$$P("1") = \lim_{n \to \infty} \frac{m}{N},$$

where $m$ represents the digital value of the storage register. It is evident that the accuracy of assessment of probability depends on $N$, and for small values of $N$ we may obtain an erroneous estimate of $P$. This error is in the form of variance [11].

The most common arithmetic operations in digital signal processing are summation and multiplication. In addition, shift operation is needed as well. Traditionally, in Stochastic Signal Processing (SSP) the product of two statistically uncorrelated signals is computed by a single AND gate. Summation is performed using an OR gate [1]. As probability of "high" levels increases, pulse overlap also increases and summation saturates

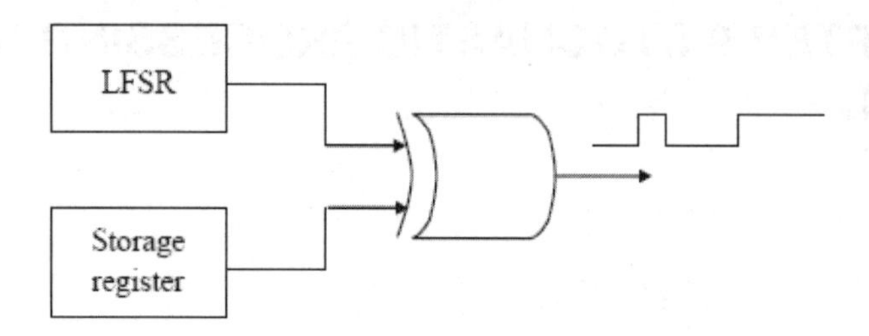


**Fig. 9.1.** Digital-to-stochastic converter

gradually. Other techniques to perform summation are described in [2]. Some authors [3] are using AND and OR gates for multiplication and addition of two delta-modulated pulse sequences. For higher input levels of an analog signal to the delta modulator this approach could be used. However, for lower levels (less than 0.2V) this approach is not justified [4]. In this chapter, we introduce a novel type of circuit for arithmetic operations on a stochastic delta-sigma modulated pulse density stream. First, we will briefly repeat the basic idea of digital implementation of continuous time filters using a stochastic approach, presented in [1]. Second stochastic delta-sigma modulator (SΔ-ΣM) is presented. Finally, a universal arithmetic unit is introduced for processing the SΔ-ΣM pulse stream. As an example, low-pass and high–pass filters are simulated.

## 9.2 EXISTING APPROACH

Authors of [1] exploited the similarity between probability and Boolean algebra to obtain simple and inexpensive realization of stochastic filters. The block diagram of a first order low-pass filter is shown in fig. 9.2. The input signal is added with level "1" using a wired OR summation. This module implements the following equation

$$Add(X_k, Y_k) = X_k \oplus Y_k \quad + \quad X_k \overline{(sig(X_k) \oplus sig(Y_k))}\,.$$

The output of the OR summation circuit is a low-density error signal. This error is time integrated with an up/down counter and then stochastically converted according to Fig. 9.2. The output signal is a stochastic pulse sequence that follows the input pulse stream. The dynamics of this sequence depends on the counter size and the clock frequency. The time constant and the cutoff frequency of the proposed system is

$$\tau = \frac{2^n}{f_{clk}}, \quad f_{clk} = \frac{1}{2\pi\tau}$$

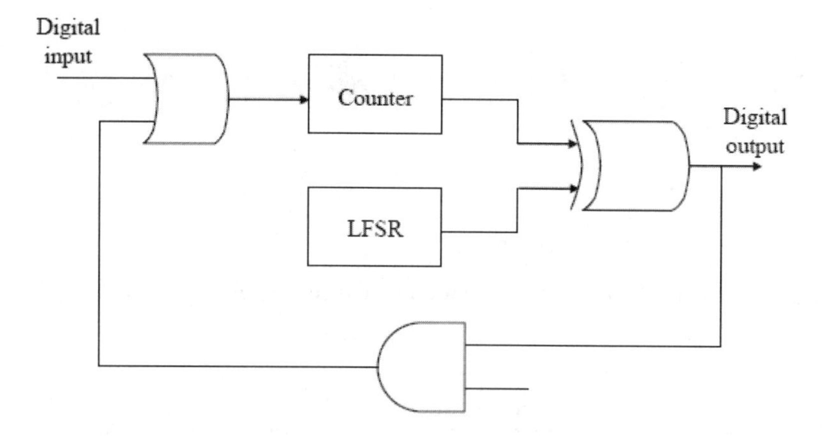


**Fig. 9.2.** First order low-pass filter [1]

where $n$ is the counter size, and $f_{clk}$ is the system clock frequency. The system transfer function is then given by

$$G(s) = \frac{1}{1 + s\tau}.$$

To achieve a $1/k$ gain factor, $k$ is introduced in the feed-back loop. In this case, the transfer function is defined as

$$G(s) = \frac{1/k}{1 + s\tau/k}.$$

The proposed high-pass filter implementation is shown in fig. 9.4 [1]. It is implemented with a LPF whose output is inverted and multiplexed with the input. The high-pass filter transfer function is given by

$$G(s) = \frac{s\tau}{1 + s\tau}.$$

With the gain factor k, the new transfer function is

$$G(s) = \frac{(k-1) + s\tau}{k + s\tau}.$$

This approach, however, has limitations because of saturation. This can be a serious problem for higher levels of an input signal. Special care must be taken to avoid saturation, which is often not simple. This problem can be avoided using a stochastic delta-sigma modulator as an A/D converter and using special proposed circuits for arithmetic operation. In the following sections, we will show that the same circuit can be implemented using this approach.

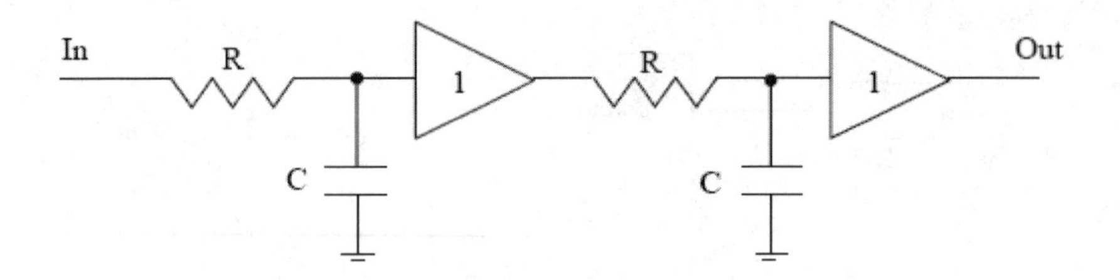


Fig. 9.3. Equivalent analog system with impedance isolation

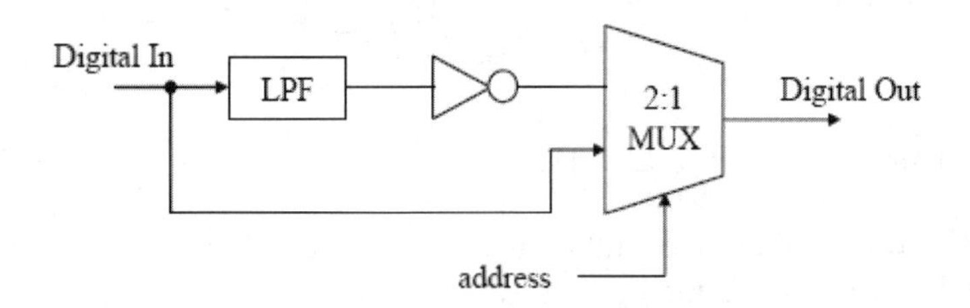


Fig. 9.4. High pass filter scheme [1]

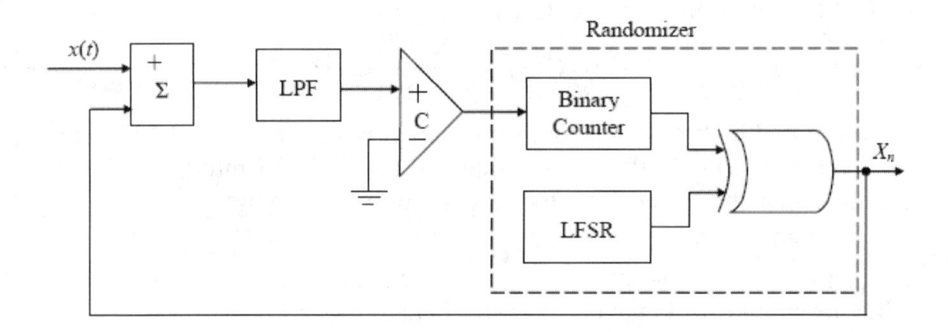


Fig. 9.5. Stochastic Delta-Sigma Modulator

## 9.3 STOCHASTIC Δ-ΣM ANALOG-TO-DIGITAL CONVERTER

Linear delta modulation (LΔM) and delta-sigma modulation are well understood in the literature (see chaps 5 & 6). The stochastic analog-to-digital delta-sigma modulator is introduced in [7]. The structure of this SΔ-

ΣM consists of ordinary delta-sigma modulator with an embedded stochastic low-pass filter in the forward path. This low-pass filter uses OR gate for summation. The digital output is decimated.

Next we propose a SΔ-ΣM implementation as in Fig. 9.5. We see that SΔ-ΣM consists of identical elements as ordinary Δ-ΣM, except circuitry for randomization. One possible implementation of SΔ-ΣM is to use FPGA [7].

## 9.4 UNIVERSAL Δ-ΣM ARITHMETIC UNIT

Arithmetic operations on a linear delta-modulated pulse stream were introduced by Kouvaras [8] and others. Zrilic [9] proposed a novel type of universal arithmetic unit for ternary delta-modulated pulse stream. Freedman and Zrilic [10] extended previous work and have shown that in addition to linear arithmetic operations on a delta-modulated pulse stream, non linear operations are possible as well. Implementation of this algorithm is fully disclosed in [4]. $E_{n+1} = E_n + f(.) - L\text{sgn}(E_n)$, where $E_{n+1}$ and $E_n$ are present and the previous value of the digital signal at the output of the modulator, $f(.)$ function to be implemented. $L$ is a constant whose value is dependent on the length of the shift register and the function to be implemented [4,10]. Fig. 9.6 illustrates the case of the multiplication and summation of two input signals, when a delta-sigma arithmetic unit (DSAU) is used.

It is worth mentioning that this circuit performs well for all levels of input signals and does not suffer from saturation as the OR gate [1]. The same circuit performs all operations and the value of the constant $L$ is different depending on the operation performed.

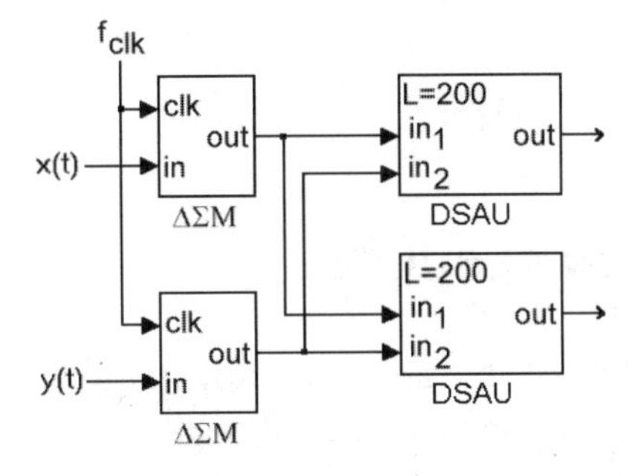


*(a)*

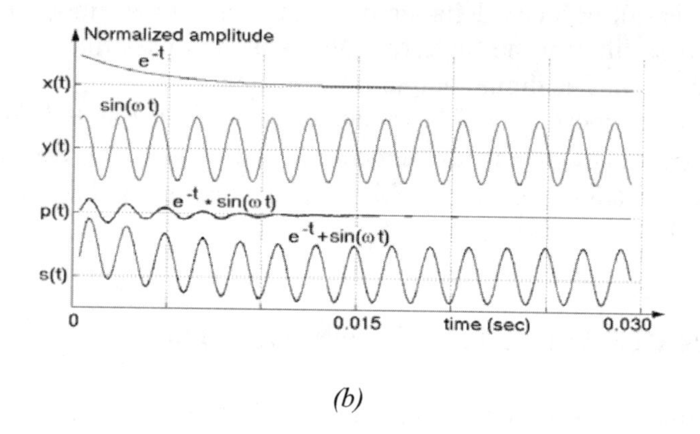


*(b)*

**Fig. 9.6.** Illustration of multiplication and addition of two delta-sigma modulated sequences $X_n$, $Y_n$ [12]

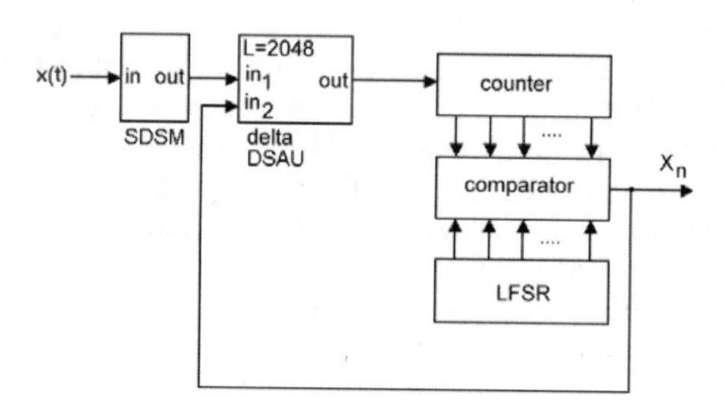


**Fig. 9.7.** First-order LPF [12]

## 9.5 SIMULATION RESULTS

To illustrate validity of our approach, low-pass and high-pass filters are simulated using SΔ-ΣM and a universal delta DSP arithmetic unit. Fig. 9.7 illustrates the block diagram of the low-pass filter proposed. The cut-off frequency of the low-pass filter is $f_c$ = 180.00 Hz, the sampling frequency $f_s$ = 512 kHz and the length of LFSR is $2^{10}$. Fig. 9.8 presents a frequency response of the first order LPF from Fig. 9.7.

A high pass filter can be constructed as shown in Fig. 9.9. The cut-off frequency of this filter is $f_c$ = 110.00 Hz and thc sampling frequency $f_s$ = 512 KHz. Fig. 9.10 presents the frequency characteristic of this filter.

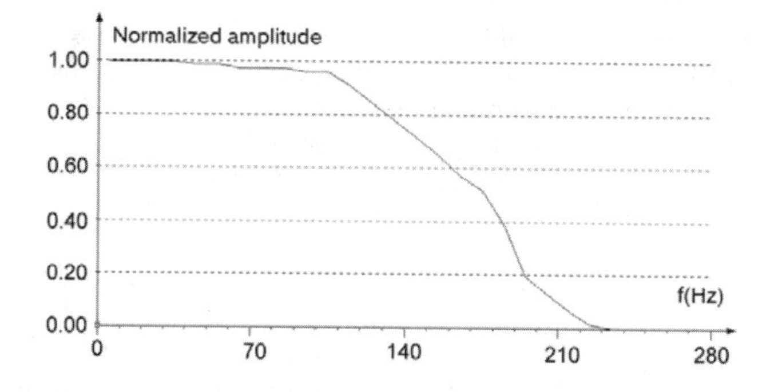


**Fig. 9.8.** Frequency response of the first-order LPF [12]

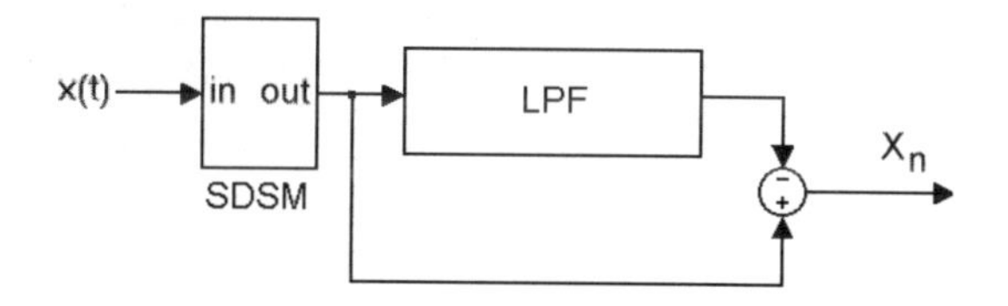


**Fig. 9.9.** High-pass filter scheme [12]

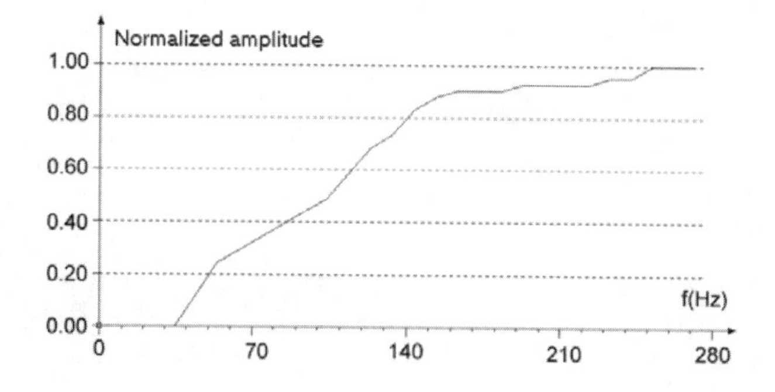


**Fig. 9.10.** Frequency response of the first-order HPF [12]

## 9.6 CONCLUSION

In conclusion, we have shown that the same performance can be achieved as in [1]. In addition, our approach does not suffer from saturation problems and the level of the input signal of SΔ-ΣM can change from zero to supplied voltage. This approach to filter design allows a low area cost of implementation in programmable devices.

## REFERENCES


1. J. M. Quero et. al., "Continuous Time Filter Design Using Stochastic Logic", *Proceedings of 42$^{nd}$ Midwest Symposium on Circuits and Systems*, Vol. 1, pp. 113-116.
2. C. L. Janer, J.M. Quero, L.G. Franquelo, "Fully Parallel Summation in a New Stochastic Neural Network Architecture", *IEEE International Conference on Neural Networks*, pp. 1498-1503, San Francisco, 1993.
3. C.J. Kikkert, "Amplitude and Frequency Modulators Using a Switchable Component Controlled by Data Signals", US patent #4,320,361.
4. Dj. Zrilic, "Circuits and Systems for Functional Processing of Delta Modulated Pulse Density Stream", US patent pending.
5. R. Steel, "Delta Modulation Systems", Pentech Press, London, 1975.
6. J. Candy, G. Temes, "Oversampling Delta-Sigma Data Converters", IEEE Press, 1992.
7. S. L. Toral, "Stochastic A/D Sigma-Delta Converter on FPGA", *Proceedings of 42$^{nd}$ Midwest Symposium on Circuits and Systems*, Vol. 1, pp.35-38.
8. N. Kouvaras, "Operations on Delta Modulated Signals and their Applications in the Realization of Digital Filters", *Institute of Electronic and Radio Engineers*, Vol. 48, No. 9, September 1978, pp.431-438.
9. Dj. Zrilic et. al., "Implementing Signal Processing Functions on Tenary Encoded Delta-Modulated Pulse Streams", *Proceedings of IEEE ICAS'88*, Helsinki, pp. 1553-1556.
10. M. Freedman, Dj. Zrilic, "Nonlinear Arithmetic Operations on the Delta-Sigma Pulse Stream", *Elsevier, Signal Processing*, 21 (1990), pp. 25-35.
11. B.R. Gaines, "Stochastic Computing Systems" *Advance in Information Science*, edited by J.T. Tou, Plenum Press, New York, 1969, pp. 37-112.
12. Dj. Zrilic, N. Pjevalica, "Stochastic Signal Processing Using Delta-Sigma Modulation", *World Automation Congress*, WAC 2002, Orlando, FL.

## REFERENCES

# CHAPTER 10 MEASUREMENTS BASED ON Δ-ΣM

## 10.1 DIRECT DYNAMIC MEASUREMENT WITH INTERVAL UNCERTAINTY

In many real-life situations, we want to monitor the value of a physical quantity $x$ for all moments of time $t$ (e.g., to check if the object of measurement is performing in the right manner). We want to make as many measurements as possible, so that we have more information to send. However, the capacity of the communication channel is limited (and in many situations, e.g., in space exploration, the cost of adding an extra communication channel can be enormous). The smaller the interval between consequent measurements the more information we need to send. Consequently, the limit on the capacity of the communication channel restricts the interval between the measurements. Let us denote the smallest time between the measurements (that the communication channel can still support) by $\Delta t$. If we denote the starting moment for our monitoring by $t_0$, then, since we want to measure as many values of the quantity $x(t)$ as possible, we will measure the value of $x(t)$ of the quantity $y$ in the moments of $t_0$, $t_1 = t_0 + \Delta t, \ldots, t_k = t_0 + k\Delta t$, etc.

Measurements are never absolutely precise [3]; therefore, the measurement result $x^*$ can differ from the actual value $x$ of the corresponding quantity by the *measurement error* $\Delta x = x^* - x$. For a measuring instrument or a sensor to make sense, the manufacturer must provide us with the guaranteed error (if there is no guaranteed error, then we can conclude nothing from the fact that, e.g., the measure value was $x^* = 10.0$; the actual value can be 9.9, can be 2,000 ...). In some cases, we know the probabilities of possible errors $\Delta x$. In many cases, however, the guaranteed upper bound $\Delta$ is the only information about the errors $\Delta x$ that the manufacturer provides. In these cases, the only information that we know about the actual value of $x$ is that this actual value belongs to an interval $x = [x^* - \Delta, x^* + \Delta]$.

## 10.2 THE MAIN IDEA BEHIND DELTA-MODULATION

A possibility to decrease the number of transmitted bits (and thus, to make more frequent monitoring measurements) comes from the fact that the measured quantities are usually changing continually, and we usually know the upper estimate $M$ on the rate with which the measured quantity $x(t)$ changes (if we do not have any limits $M$, then we have no information about the intermediate values $x(t)$, and our monitoring is of limited usage). In this case, if we know the value $x(t_k)$ in the moment of time $t_k$, then the next value $x(t_{k+1})$ cannot deviate from $x(t_k)$ by more than $M\Delta t$. Let us give an example of why this idea can indeed decrease the number of bits that is necessary to carry a single measurement.

**Example:** Let us assume that we are measuring the temperature every millisecond, with an accuracy of one degree, $|x^*(t_k) - x(t_k)| \leq \Delta = 1$. Let us also assume that the measured value $x^*(t_k)$ of the temperature $x$ at the same moment $t_k$ is equal to 1,826 degrees, and we know that during the interval between the two consequent measurements (i.e., during one millisecond) the temperature can change by no more than two degrees, i.e. $|x(t_{k+1}) - x(t_k)| \leq 2$. Therefore, the difference between the *measured* values of temperature cannot exceed four degrees, $|x^*(t_{k+1}) - x^*(t_k)| \leq |x^*(t_{k+1}) - x(t_{k+1})| + |x(t_{k+1}) - x(t_k)| + |x(t_k) - x^*(t_k)| \leq 1 + 2 + 2 = 5$.

According to the traditional approach, in the next moment of time $t_{k+1}$, we must send the numerical value of the measured temperature $x^*(t_{k+1})$. This value is an integer between $1{,}822 = 1{,}826 - 4$ and $1{,}830 = 1{,}826 + 4$. Therefore, it is an integer between $1{,}024 = 2^{10}$ and $2{,}948 = 2^{11}$, so, we need 11 binary digits to describe this measurement result. On the other hand, instead of sending the value $x^*(t_{k+1})$, we can simply send the difference between $x^*(t_{k+1})$ and $x^*(t_k)$. This difference is an integer between –4 and 4, so it has only 9 possible values (-4, -3, . . . , 0, 1, . . . ,4). We only need four bits (one bit for sending a sign, and three bits for sending the absolute value of the difference) as opposed to 11 in the traditional approach. Since we need fewer bits to send the results of the measurements, we can hold measurements 11/4 (>2) times more frequently than before.

We have already mentioned that ideally, we should be monitoring the value of $x(t)$ for every moment $t$, but in reality, we only get the values in the moments $t_1, \ldots, t_k, \ldots$ . Therefore, if we are interested in the value of $x(t)$ for some intermediate moment of time $t$, i.e., in a moment of time that lies in between $t_k$ and $t_{k+1}$ for some $k$, then as an estimate for $x(t)$ we take the latest available measured value, i.e., $x^*(t_k)$. Even if we measured $x(t_k)$ precisely, this difference in times between $t$ and $t_k$ would still contribute to an error in this estimate, an error $x(t)$ - $x(t_k)$ that is limited by $M(t - t_k) \leq M.\Delta t$. The additional measurement error $x^*(t_k)$ - $x(t_k)$ may increase the total error $x(t)$ - $x^*(t_k)$ of using $x^*(t_k)$ as an estimate for $x(t)$. Since we already have an error component of size $M\Delta t$, it makes no big sense to measure the values $x(t_k)$ with accuracy that is much better than $M\Delta t$. There is no sense in trying to achieve measurement errors that are much smaller than $M\Delta t$. Such super-accurate measurements would mean using very expensive sensors, but their usage will not seriously improve the resulting error, because this error would still be of order $M\Delta t$. Consequently, the measurement accuracy $\Delta$ is usually chosen to be smaller than $M\Delta t$, but approximately of the same order ($\Delta < M\Delta t$, $\Delta \approx M\Delta t$). With this choice, the difference $x(t_{k+1})$ - $x(t_k)$ (that is $M\Delta t$) is measured with an error that is close to the value of this difference. With such a huge measurement error, we can basically distinguish between only two cases:

- The case when this difference is positive.
- The case when this difference is negative.

As a result, the sensor gets the measurement results $x^*(t_1)$, ... but it sends only one bit per moment of time for processing. This bit actually represents a sign of the difference between the two consequent values of the signal (i.e., whether $x$ increased with respect to the previous moment of time or not), so it is natural to represent this bit not as zero or one, but as a *sign*, i.e., as +1 or –1. Let us denote the sign bit that comes out of the sensor in the moment $t_k$ by $s(k)$. Then, at the receiving end of the communication channel, all we have is a sequence of sign bits $s(1)$, ..., $s(k)$ . In order to be able to reconstruct the signal from this sequence, we must know the initial value of the signal $x^*(t_0)$. How can we reconstruct the signal from this sequence? There is not much that we can do but follow the following natural algorithm:

1. As the initial value $r(0)$ of the reconstructed signal $r$, we simply take $r(0) = x^*(t_0)$
2. To get further reconstructed values $r(k)$, we proceed as follows:
   - If we have already computed $r(k)$, and the next bit that comes out of the communication channel is 1, we add $\alpha = M\Delta t$ (i.e., take $r(k + 1) = r(k) + M\Delta t$).
   - If we have already computed $r(k)$, and the next bit that comes out of the communication channel is –1, we subtract $M\Delta t$ (i.e., take $r(k + 1) = r(k) - M\Delta t$).

This natural reconstruction algorithm leads to the following natural idea of selecting a signal $s(k)$ that would go through the communication channel:

- At every moment of time, after we generate the communication bit, we also simulate the reconstruction procedure at the sensor's end of the communication channel (thus getting $r(0)$, $r(1)$, ...).
- After a new measured value $x^*(t_k)$ arrives, we compare it with the previously reconstructed signal $r(k-1)$. Now we have two options:
  - If we choose $s = +1$ then the next reconstructed signal $r(k)$ will be greater than $r(k-1)$.
  - If we choose $s = -1$ then the next reconstructed signal $r(k)$ will be smaller than $r(k-1)$.

  So, to get $r(k)$ as close to $x(t_k)$ as possible, we will choose

  - $s(k) = +1$ if the value of $r(k-1)$ is smaller than $x^*(t_k)$ (and thus needs to be increased).
  - $s(k) = -1$ if the value of $r(k-1)$ is greater than $x^*(t_k)$ (and thus needs to be decreased).

The resulting algorithm is called *delta-modulation.*

*Comment*: For a recent survey of delta-modulation techniques, see [2] and references therein. These methods and results, however, are mainly developed for the statistical case.

## 10.3 DIRECT DYNAMIC MEASUREMENT AND ITS ERROR ESTIMATE

**Definition 1.** By a *dynamic measuring instrument*, we mean a pair $(\Delta,\Delta t)$ of two positive numbers:

- a number $\Delta > 0$ will be called the *measurement accuracy,*
- a number $\Delta t > 0$ will be called a *time quantum.*

**Definition 2.** By a *dynamic measurement situation*, we mean a set $(I, M, t_0, x, \{t_k\}, \{x^*(t_k)\})$, where

$I$ is a dynamic measuring instrument,

$M$ is a positive real number called the *(prior) bound on the rate of change of the signal.* We will assume that $\Delta \leq M\Delta t$,

$t_0$ is a real number called the *initial moment of time,*

$x$ is a function from real numbers into real numbers that is an $M$-Lipshitz function (i.e. $|x(t) - x(s)| \leq M|t - s|$ for all $t$ and $s$),

$\{t_k\}$ $0 \leq k$, is a sequence of real numbers defined as $t_k = t_0 + k.\Delta t$. A number $t_k$ will be called $k^{th}$ *measurement moment*, and

$\{x^*(t_k)\}$ is a sequence of real numbers for which for every $k$, $|x^*(t_k) - x(t_k)| \leq \Delta$. The element $x^*(t_k)$ will be called the *result of $k^{th}$ measurement.*

*Comment*: Let us first consider the case when we do not use delta-modulation

**Definition 3**. For every dynamic measurement situation, and for every moment of time $t \geq t_0$, by a *monitoring error*, we mean the difference $x^*(t_k) - x(t)$, where $k$ is the largest value for which $t_k \leq t$.

**Proposition 1**

- For every dynamic measurement situation, and for every moment of time $t$, the absolute value of the monitoring error does not exceed $M\Delta t + \Delta$.
- For every $\delta$, there exists a dynamic measurement situation and a moment of time $t \geq t_0$ for which the monitoring error is not smaller than $M\Delta t + \Delta - \delta$.

*Comments:*

1. These two statements mean that $M\Delta t + \Delta$ is the error bound for monitoring error, and no better bound is possible.
2. The fact that the error bound is $M\Delta t + \Delta$ can be easily explained by the fact that we have two sources of error:
   - The measurement error, whose bound is $\Delta$.
   - The error caused by the difference between $t$ and $t_k$ whose upper bound is $M\Delta t$.

*Proof of Proposition 1* Let us first prove that the monitoring error is always bounded by $M\Delta t + \Delta$. Indeed, if $t_k \leq t < t_{k+1}$, then $0 \leq t - t_k < t_{k+1} - t_k = \Delta t$, and therefore, by the definition of a measurement situation, $|x(t) - x(t_k)| \leq M|t - t_k| \leq M\Delta t$. Therefore, $|x(t) - x^*(t_k)| \leq |x(t) - x(t_k)| + |x(t) - x^*(t_k)| \leq M\Delta t + \Delta$. Consequently, the monitoring error is indeed always bounded by $M\Delta t + \Delta$.

Let us now show that a smaller bound for a monitoring error is impossible. Indeed, assume that $\delta > 0$, $M$, $t_0$, $\Delta t$ are fixed. As a measured signal, let us take the function $x(t) = M(t - t_0)$. As measurement results, we will take $x^*(t_k) = x(t_k) - \Delta$. Then, for $t = t_1 - \delta/M$, the monitoring error is equal to $x(t) - x^*(t_0) = M(t_1 - \delta/M - t_0) - (M(t_0 - t_0) - \Delta) = M\Delta t + \Delta - \delta$. Q.E.D.

## 10.4 DELTA MODULATION: FORMAL DEFINITION

Let us now define delta-modulation and show that its usage (while saving on communication) does not decrease the resultant monitoring error. Specifically, we will show that with delta-modulation, we can achieve the same monitoring error if we double the measuring rate. At this rate, we need to transmit twice as many measurement results. However, since we only need one bit to transmit a single delta-modulated measurement result, and we need several bits to transmit the actual measurement result $x^*(t_k)$, the resulting total amount of bits per second that needs to be transmitted is smaller when we use delta-modulation.

**Definition 4.** For every dynamic measurement situation, by the *result of delta-modulation* applied to the sequence $x^*(t_k)$, we mean the sequence $s(1), \ldots, s(k), \ldots$ whose elements are determined by the formula

$$s(k) = \text{sgn}[x^*(t_k) - (x^*(t_0) + M\Delta t \sum_{j=1}^{k-1} s(j))], \tag{10.1}$$

where

$$\text{sgn}(a) = \begin{cases} +1, & a \geq 0 \\ -1, & a < 0. \end{cases}$$

For every dynamic measurement situation by the *reconstructed* or *delta-demodulated* signal, we mean a sequence

$$r(k) = x^*(t_0) + M\Delta t \sum_{j=1}^{k-1} s(j). \tag{10.2}$$

**Definition 5.** For every dynamic measurement situation, and for every moment of time $t$, by a *monitoring error after (delta-) demodulation*, we mean the difference $r(k) - x(t)$, where $k$ is the largest value for which $t_k \leq t$.

**Proposition 2.**

- For every dynamic measurement situation, and for every moment of time $t$, the absolute value of the monitoring error after delta-demodulation does not exceed $2M\Delta t + \Delta$.
- For every $M > 0$, $t_0$, $\Delta t > 0$, $\Delta > 0$, and $\delta > 0$, there exists a dynamic measurement situation and a moment of time $t \geq t_0$, for which the monitoring error after delta-demodulation is not smaller than $2M\Delta t + \Delta - \delta$.

*Comments:*

1. These two statements mean that $M\Delta t + \Delta$ is the error bound for monitoring error after delta-demodulation, and that no better bound is possible.
2. Due to Proposition 2 if, for measurements with delta-modulation, we take the time quantum $\Delta t$ that is twice smaller than the one that was used for regular measurements; we will get exactly the same monitoring error as for measurements without modulation. Let us give two examples:
3. Suppose that we measure $x(t_k)$ with an accuracy of 1%. This means that possible measurement results run from –100 to 100 range. The binary representation of 100 takes seven bits, consequently, with an extra bit for sign, we need eight bits to transmit the result of a single measurement. If we use delta-modulation, then we only need one bit per measurement, but these measurements must be two times more frequent. When initially we needed eight bits, we now need only two. Therefore, if we use delta-modulation, we can keep the same total error and reduce the information flow by a factor of four.
4. Suppose now that we measure $x(t_k)$ with an accuracy of 0.1%. In this case, possible measurement results run from –1000 to 1000. The binary representation of 1000 takes ten bits, so we need eleven bits to transmit the result of a single measurement. If we use delta-modulation, and aim at the same accuracy of the final result, we thus need two bits during the same time quantum $\Delta t$. As a result, we decrease the information flow by a factor of 5.5.

In general, the more accurate the measurements, the more we save by using delta-modulation.

*Proof of Proposition 2* Let us first start with proving the inequality, and then produce an example that proves the second part of this proposition. To prove the inequality, we will first prove (by induction over $k$) that $|x(t_k) - r(k)| \le M\Delta t + \Delta$ for all $k$.

*Induction base* - The initial reconstructed value $r(0)$ is defined as $r(0) = x^*(t_0)$, but by definition of a measurement situation, we have $|x(t_k) - x(t_0)|$. Therefore, $|r(0) - x(t_0)| \le \Delta < M\Delta t + \Delta$.

*Induction step* - Assume that we have already proved the desired inequality for $k$, i.e.

$$|x(t_k) - r(k)| \le M\Delta t + \Delta. \tag{10.3}$$

We must prove that a similar inequality holds for $k + 1$, i.e.

$$|x(t_{k+1}) - r(k+1)| \le M\Delta t + \Delta. \tag{10.4}$$

To prove that, we will consider two cases:
1. the case when $x^*(t_{k+1}) \geq r(k)$, and therefore, $s(k+1) = 1$.
2. the case when $x^*(t_{k+1}) < r(k)$, and therefore, $s(k+1) = -1$.
In the first case, $r(k+1) = r(k) + M\Delta t$. From (10.3) we conclude that

$$x(t_k) \leq r(k) + M\Delta t + \Delta. \tag{10.5}$$

From the definition of a measuring situation, we conclude that $x(t_{k+1}) - x(t_k) \leq M(t_{k+1} - t_k) = M\Delta t$, so $x(t_{k+1}) \leq x(t_k) + M\Delta t$. Replacing $x(t_k)$ by its upper bound taken from (10.5), we conclude that $x(t_{k+1}) \leq r(k) + M\Delta t + \Delta + M\Delta t$. Since $r(k) + M\Delta t = r(k+1)$, we conclude that $x(t_{k+1}) \leq r(k+1) + M\Delta t + \Delta$, i.e. $x(t_{k+1}) - r(k+1) \leq M\Delta t + \Delta$. This is half of the desired inequality (10.4).

To complete the proof for this case, it is thus necessary to prove the other half of this inequality, i.e., to prove that $x(t_{k+1}) \geq r(k+1) - (M\Delta t + \Delta)$. If in the case under consideration, $r(k+1) = r(k) + M\Delta t$, this inequality is equivalent to $x(t_{k+1}) \geq r(k) - \Delta$. This follows from the following sequence of inequalities, in this case, $x(t_{k+1}) \geq r(k)$, by definition of a measurement situation, $x(t_{k+1}) \geq x^*(t_{k+1}) - \Delta$, and therefore, $x(t_{k+1}) \geq x^*( t_{k+1} ) - \Delta \geq r(k) - \Delta$. This inequality is thus proved, and so (10.4) is true in the first case. The second case is proved similarly. Now, the desired inequality follows in a manner similar to the proof of Proposition 1: if $t_k \leq t < t_{k+1}$ then $|x(t) - r(k)| \leq |x(t) - x(t_k)| + |x(t_k) - r(k)| \leq M|t - t_k| + M\Delta t + \Delta \leq M\Delta t + M\Delta t + \Delta$.

Let us now give an example of the measurement situation in which the monitoring error is not smaller than $2M\Delta t + \Delta - \delta$

$x$: we choose the following function $x(t)$
$x(t_2) = -M\Delta t;$
$x(t_k) = 0$ for $k \neq 2$;
$x(t)$ is linear in the intervals $[t_k, t_{k+1}]$

$x^*$: we take $x^*(t_k) = x(t_k) + \Delta$ for all $k$

$t$: we take $t = t_2 - \delta/M$

In this case, $r(0) = x^*(t_0) = \Delta$. Since $x^*(t_1) = \Delta \geq r(0)$, we have $s(1) = +1$ and $r(1) = r(0) + M\Delta t =$ c, we have $t_1 \leq t < t_2$, and since $x$ is linear on $[t_1\ t_2]$, we have $x(t) = x(t_1)(t - t_1)/(t_2 - t_1) + x(t_2)(t_2 - t)/(t_2 - t_1) = (-M\Delta t)[(\Delta t - \delta/M)/\Delta t] = -M\Delta t + \delta$. So here, $|r(1) - x(t)| = |(M\Delta t + \Delta) - (-M\Delta t + \delta)| = 2M\Delta t + \Delta - \delta$. Q.E.D.

## 10.5 FREQUENCY DEVIATION MEASUREMENT BASED ON Δ-ΣM

### 10.5.1 Problem Statement

There is need for accurate measurement of frequency deviations in many applications. For example, a frequency deviation measurement is needed in the design of power system stabilizers, power system monitors, communication systems, etc. A number of circuits, which can accurately measure the frequency deviation, are proposed in the literature. As an introduction, we will briefly review two methods related to our work.

The first method, considered in [4], is based on the multiplication of two incoming frequencies by a large factor. A BCD up/down counter is used to count a train of pulses in the up mode and a second train of pulses in the down mode. Whatever is left stored in the up/down counter is the difference between the two pulse trains. Since a frequency measurement is a pulse counting process, it was concluded that this technique could be used to find the difference between two frequencies. This measurement requires two gate-time intervals. During the first gate-time interval, the pulses of the first frequency are counted in the down mode. The counter content at the end of the two gate-time intervals is the difference between the two frequencies. To achieve a resolution of $n$ decimal places, both incoming frequencies are multiplied by the factor $10^n$. This means that the difference in the frequencies is also multiplied by the same factor. Based on this method, a digital frequency-meter was constructed and tested. It covers the range from 5 Hz to 100 Hz and provides a measurement resolution of three decimal digits. Fig. 10.1 shows a block diagram of the frequency difference meter proposed in [4].

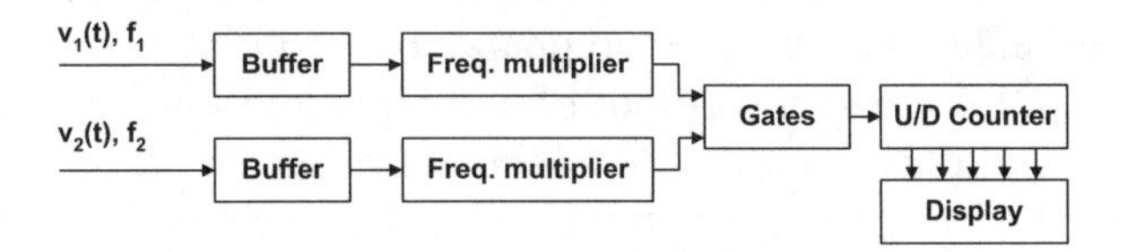


**Fig. 10.1.** Block diagram of a frequency difference meter [1]

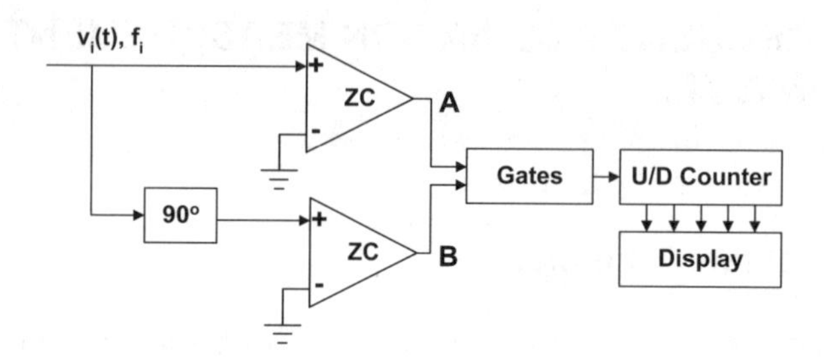


**Fig. 10.2.** The electronic bridge approach [5]

The sinusoidal input signal, for which the frequency deviation measurement is required, is passed through a zero-crossing detector to convert it into pulse train *A*. The input signal is also applicd to a phasc shifting circuit of 90 degrees. The phase-shifted signal is then converted into pulse train *B* passing it through a zero crossing detector. A two arm bridge is created and the outputs of the AND gates are pulse trains with an equal mark/space ratio. In this case, the bridge is balanced and the counter is in the zero position. When the phase shift is different from 90 degrees, the mark space ratio of one pulse train becomes higher, and for the other pulse train becomes less. As a result, the bridge is not balanced and the counter will count up or down.

In [5] a binary-coded decimal up/down counter is used to find the difference in pulse count, which is an indication of frequency deviation. In [6] a power system stabilizer sensing frequency deviation meter has been developed. A special type of frequency transducer, based on sample and hold principles, was used. However, our work is closely related to work found in [5].

Although the reference [5] approach is simple, it poses several inherent problems. First, in many practical situations, unwanted voltage fluctuations (noise) appear on the reference terminal of a comparator. This noisy reference voltage may cause a comparator to erratically switch the output state. Fig. 10.3 shows the output of a comparator when a noise signal (of the variance of only one promile of sine wave amplitude at the input of the comparator) is superimposed on the zero reference voltage.

Second, unwanted noise is frequently superimposed on the input signal as well. Fig. 10.4 shows the case when a noisy sine wave is applied to the input of a comparator with zero threshold.

When the sine wave amplitude approaches zero, the fluctuations due to noise cause the total input to vary above and below zero several times, thus producing an erratic output. In order to make the comparator less sensitive to noise, a Schmitt trigger circuit may be employed. This possibility was not suggested in reference [5]. Unfortunately, a comparator with hysteresis (the Schmitt trigger) does not solve the problem completely.

Here, we describe a digital frequency deviation meter, which is based on use of delta-sigma modulation (Δ–ΣM) and use of arithmetic operations on its pulse density stream. First, basic operations of Δ–ΣM will be described, and then the possibility of adding of delta modulated sequences will be introduced. Simulation results of the proposed method will be contrasted and compared with the method proposed in reference [5].

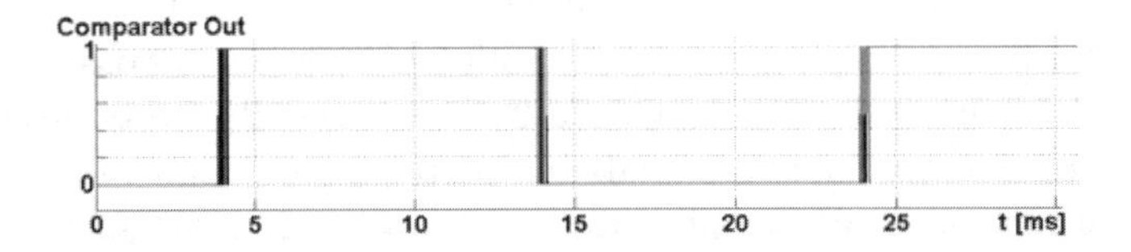


**Fig. 10.3.** Effect of noisy threshold on comparator circuit [10]

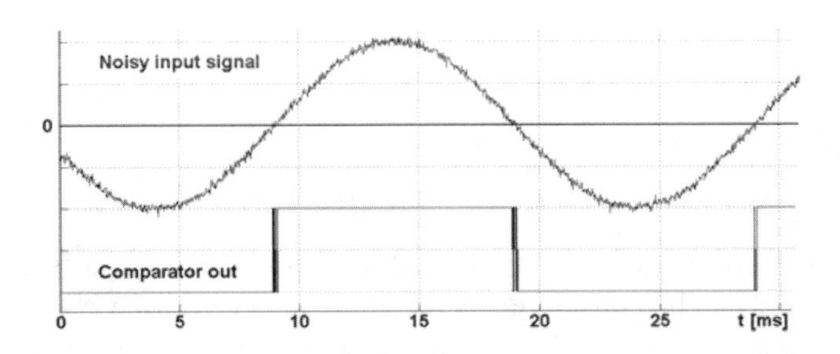


**Fig. 10.4.** Effect of noisy input signal on comparator circuit [10]

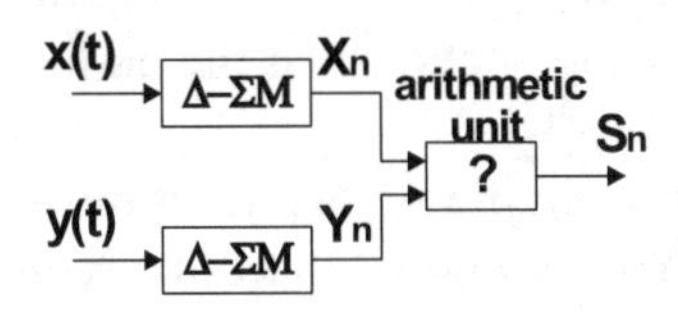


**Fig. 10.5.** System for arithmetic processing of two Δ–ΣM sequences [10]

### 10.5.2 Addition of Δ–ΣM Signals

Suppose that we have a well designed and highly over-sampled Δ–Σ modulator ($f_s >> 2 f_{in}$). The question now is if we can perform direct arithmetic operations on a serial Δ–Σ pulse stream, as shown in fig. 10.5.

The work of Kouvaras [9] presents a full theoretical treatment of the complete addition and subtraction of a binary delta modulated pulse stream. The method proposed by Kouvaras also provides information concerning the errors introduced by the operations. Although Kouvaras has analyzed the addition of linear delta modulated (LΔM) signals, Zrilic [10] has shown that the addition of Δ–Σ modulated signals is possible as well. The same binary full adder proposed by Kouvaras [9] can be used for the addition of Δ–Σ modulated signals by interchanging roles of the sum and carry out terminals of the full adder. According to Kouvaras, the newly derived sum of two synchronous delta modulated sequences is defined as

$$S_n = 0.5\{X_n + Y_n - (1 - X_n Y_n)C_n] \quad (10.6)$$

$$C_n = X_n Y_n C_{n-1} \quad (10.7)$$

$$C_{n-1} = +1 \text{ or } -1,\ n = \ldots, -1, 0, +1, \ldots . \quad (10.8)$$

The terms of $S_n$ take the values of +1 or –1. After demodulation (low pass filtering) of $S_n$, one can get

$$s(t) = 0.5[x(t) + y(t)] - 0.5[e_1(t) + e_2(t)] + \varphi(t). \quad (10.9)$$

where $0.5[e_1(t) + e_2(t)]$ is the half-sum of quantization errors of two Δ–Σ systems and can be considered as the equivalent error of a Δ–Σ system, the input of which is the analog signal $0.5[x(t) + y(t)]$. Kouvaras has shown that the error $\varphi(t)$, due to the introduction of a binary full adder, can be considered negligible. This error decreases and becomes more negligible if the step size of the linear delta modulator decreases and the sampling frequency is correspondingly increased. Using an identical delta adder, we will show in the following section that the addition of two Δ–Σ sequences is possible as well.

Fig. 10.6 shows an example of the addition of two Δ–Σ modulated sequences $X_n$ and $Y_n$. After demodulation (averaging) of $S_n$, one half of the sum is obtained. For this example, $x(t) = \sin(\omega t)$ and $y(t) = e^{-t}\sin(\omega t)$, where $f_{in} = 50$ Hz and $f_s = 100$KHz.

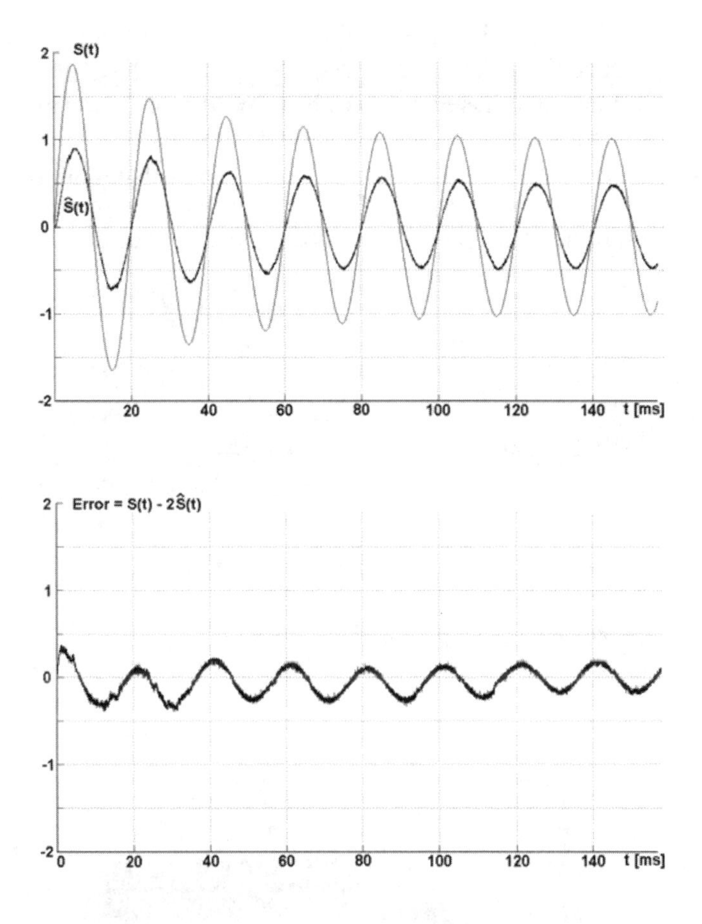


**Fig. 10.6.** An example of addition of two Δ–Σ sequences [10]

As expected, the resulting signal is half the amplitude of the sum of the input signals. The error signals can be made smaller with an increase in sampling frequency.

## 10.5.3 Implementation Method

Fig. 10.7 shows the block diagram of the proposed system for frequency deviation measurement.

The proposed system consists of two synchronous first order Δ–Σ modulators, a conventional binary full adder and an up and down counter with display. The up and down counter is a demodulator and plays the role of averager. The sinusoidal signal, for which the frequency deviation measurement is required, is passed through two synchronous Δ–Σ modulators that produce the pulse streams sequences $X_n$ and $Y_n$. The signal $y(t)$ is phase shifted and its phase angle is given by $\Phi = \pi - 2arctg(\omega RC)$. The phase shift is adjusted to 90 degrees, and in this case the output sequence $S_n$ is equally spaced with amplitude +1 or –1. After demodulation (averaging), the value of the sum is zero. Fig. 10.8 presents this case. We can see that a certain initial time is needed to settle the output of the averager.

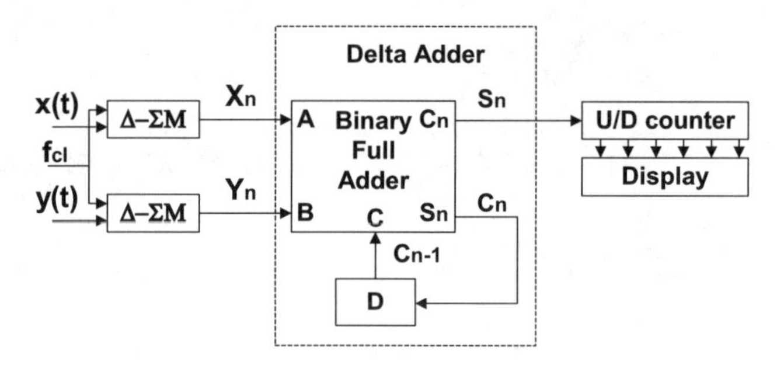


**Fig. 10.7.** Block diagram of a proposed instrument [12]

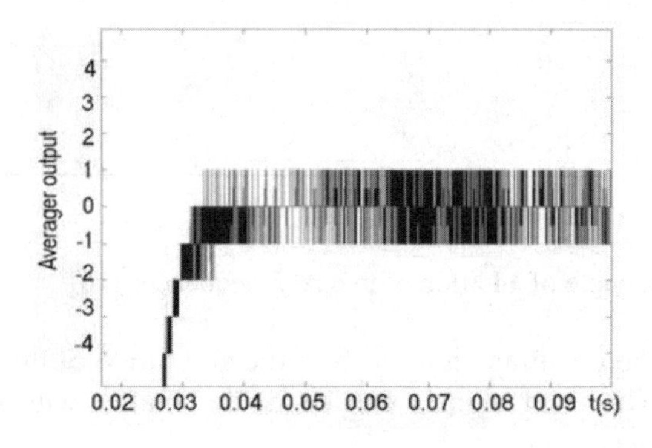


**Fig. 10.8.** Output of the averager when input frequency is 50 Hz, balanced bridge

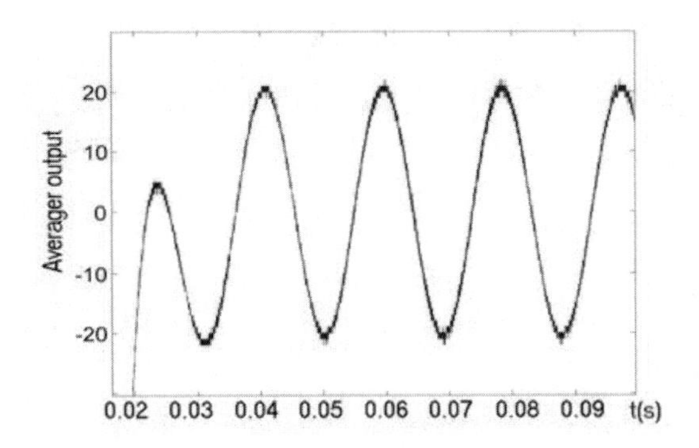


**Fig. 10.9.** Output of the averager when the bridge is out of balance for fin=52.5 Hz, (5%) [10]

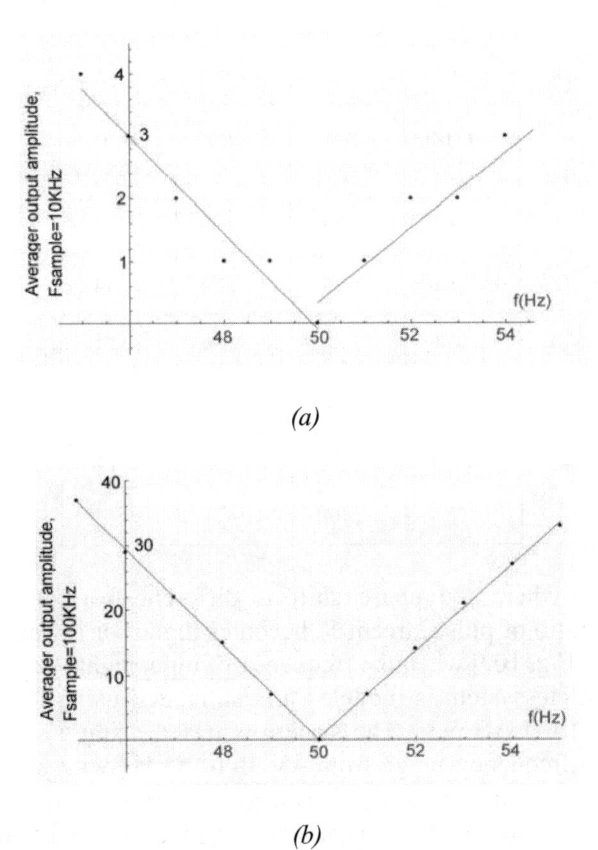

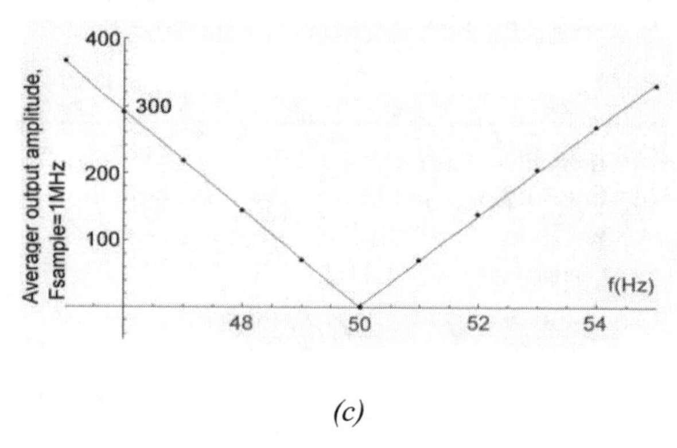


*(c)*

**Fig. 10.10.** a, b and c: Averager output amplitude as a function of pulse counts for three different sampling frequencies [10]

**Table 10.1.**

| $f_{in}$ (Hz) | Amp. output | $f_{in}$ (Hz) | Amp. output |
|---|---|---|---|
| 45.0 | 37 | 51.0 | 7 |
| 46.0 | 29 | 52.0 | 14 |
| 47.0 | 22 | 53.0 | 21 |
| 48.0 | 15 | 54.0 | 27 |
| 49.0 | 7 | 55.0 | 33 |
| 50.0 | 0 | | |

In a case where the phase shift is different from 90 degrees, the mark/space ratio of pulse stream $S_n$ becomes higher or lower. This case is illustrated in Fig. 10.9 when the frequency of input signal deviates by 5%.

The complete system is modeled and simulated using the SIMULINK toolbox in MATLAB 5.3. The system is tested using a sinusoidal input signal in the frequency range from 45 Hz to 55 Hz with steps of 1.0 Hz. Figs. 10.10 a, b, and c show the cases when the sampling frequency of Δ–Σ modulators is set to 10 KHz, 100 KHz and 1 MHz, respectively. Table 1.1 shows the numerical results presented in fig. 10.10b. As expected, we can conclude that by increasing the sampling frequency, the accuracy of measurement is increased as well. According to Kouvaras [9], both modulators must be synchronous. Figs. 11 a, b, and c show the cases of simulation for a sampling frequency of 100 KHz. An offset of 1%, 5% and 10% respectively, is introduced in the lower Δ–ΣM of Fig. 10.7. The linear interpolated line is obtained using the least-squares error method. From these results, we can conclude, that for accurate measurement, synchronism of both Δ–ΣMs must be achieved. In addition, we have simulated different component mismatch scenarios. For example, a mismatch of the cut-off frequencies of Δ–ΣM integrators does not have any influence on linearity. This is because Δ–ΣM converters are especially insensitive to circuit imperfections and component mismatch. They employ only a simple two-level quantizer, and that quantizer is embedded within a feedback loop.

### 10.5.4 Performance Comparison

For comparison, fig. 10.12 presents a simulation block diagram for both [6] and the newly proposed method. To keep the block diagram simple, the noise generators for the comparators and BQ are not shown. To compare the sensitivity of the proposed system, suggested in reference [6], to the delta-sigma approach, a noise signal of variance 1% of the input signal amplitude is added to the threshold of the BQ of the Δ–ΣM.

Fig. 10.13 shows digital outputs of the Δ–ΣM with and without noisy threshold. Due to noise added to the threshold of BQ, initial conditions of the delta-sigma system are different, thus the output pulse stream is different, but spike pulses are not present. After demodulation of $X_n$, with and without a noisy threshold, we get the same result. We can see that 1% of noise does not have any effect on performance of the Δ–ΣM, whereas one profile of noise added to the threshold of the comparator [6] has a catastrophic influence. The benefit of negative feedback of Δ–ΣM is crucial in minimization of the error caused by both mismatch and induced noise.

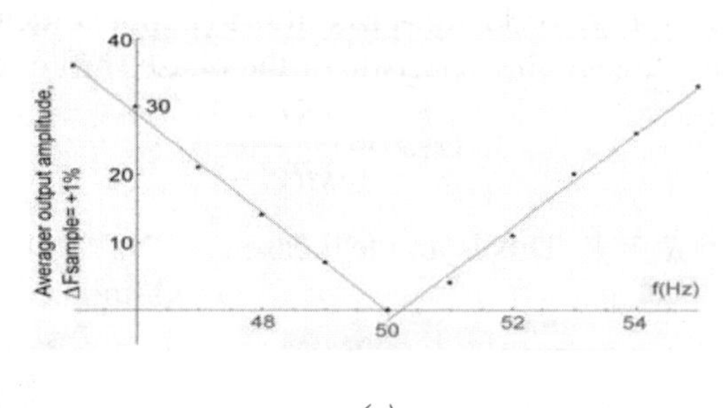


*(a)*

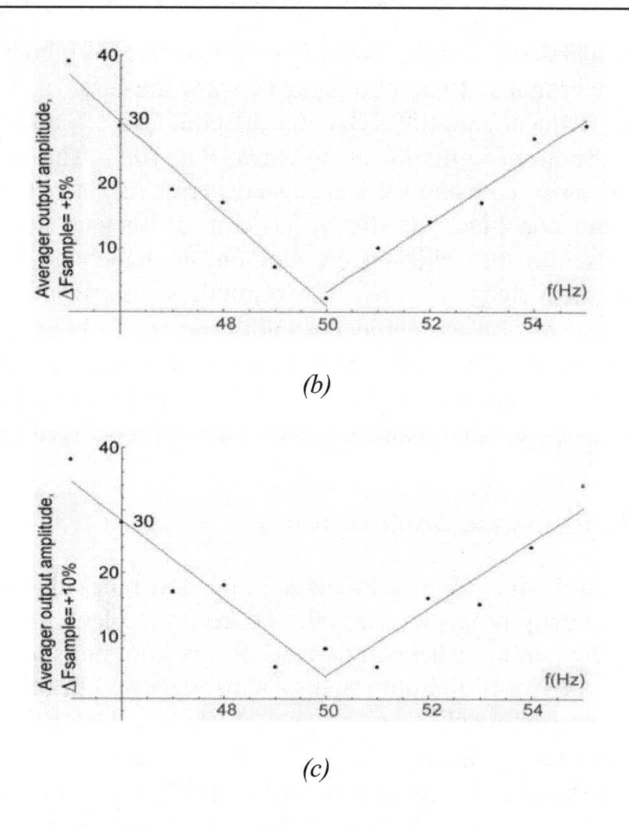


**Fig. 10.11.** a, b, and c.: Averager output amplitude as a function of pulse count for three different frequency offsets of the lower Δ–ΣM in Fig. 10.7 [10]

Fig. 10.14 shows the output of the Δ–ΣM when a noise variance of 1% of the input signal amplitude is superimposed on the input signal.

Again, we can see that both outputs are without noise spikes, which was not the case of the comparator solution (Fig. 10.4). It is worth mentioning that the sensitivity of both systems to changes in the components of the phase shift circuit, Fig. 10.15 (all-pass filter) is almost identical. It is easy to show that the transfer characteristic of the phase shift circuit is

$$G(s) = \frac{sRC - 1}{sRC + 1},$$

where $R_1 = R_2 = R_3 = R$. This is an ideal case.

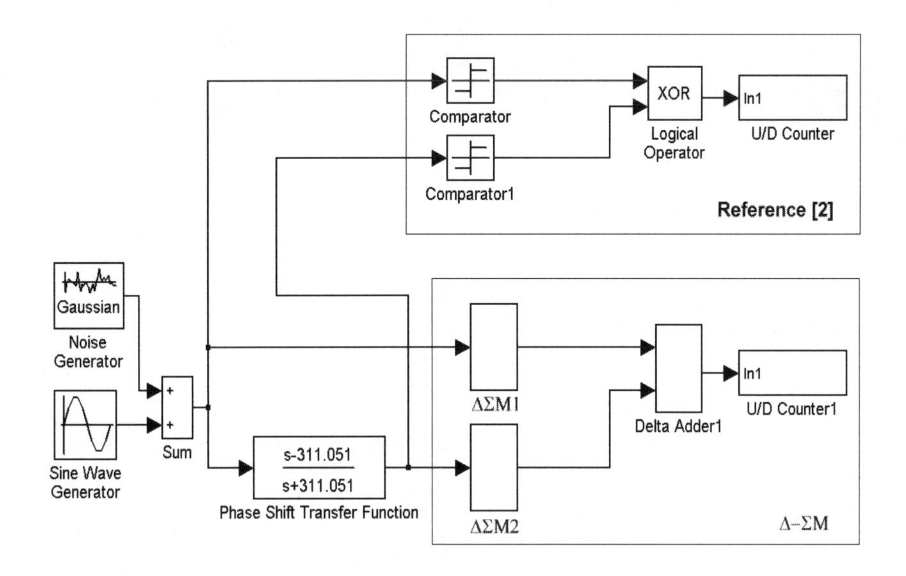


**Fig. 10.12.** Simulation block diagram [10]

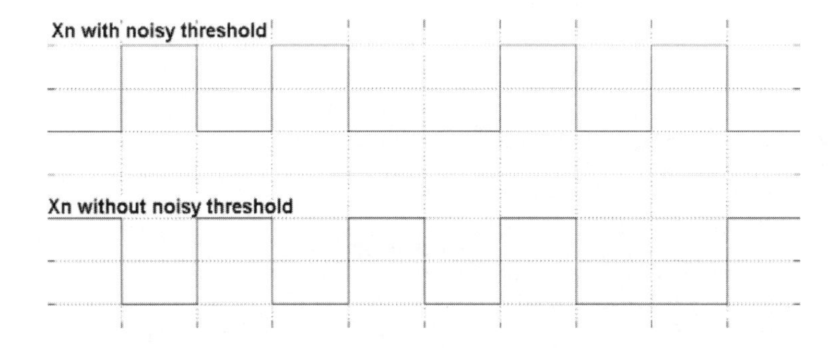


**Fig. 10.13.** Output $X_n$ of Δ–ΣM with and without noisy threshold [10]

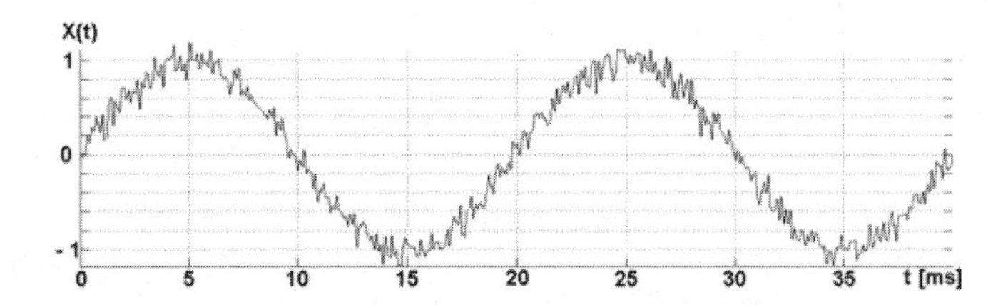

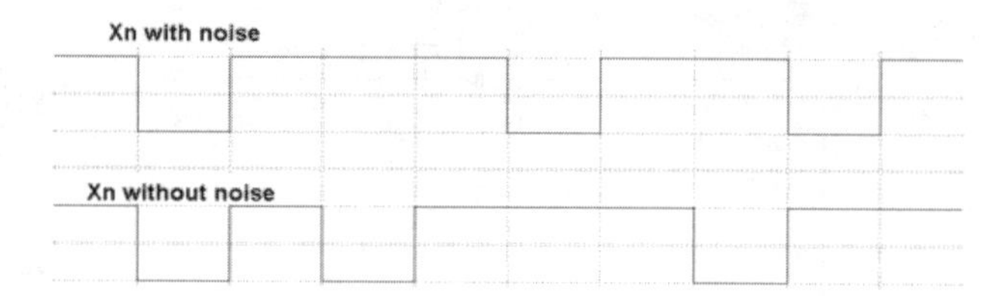


**Fig. 10.14.** Influence of noisy input [10]

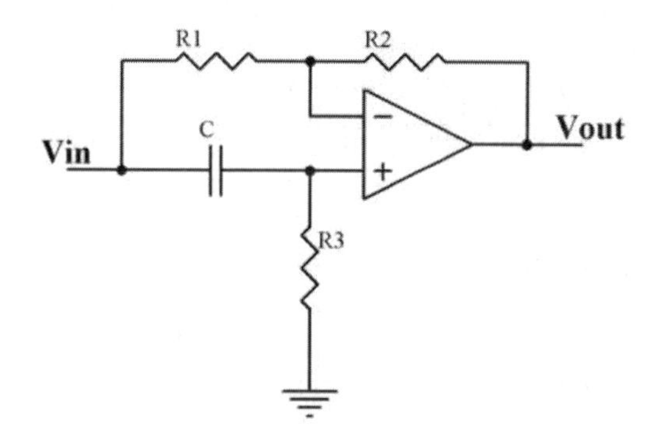


**Fig. 10.15.** Phase shift circuit used in both methods [10]

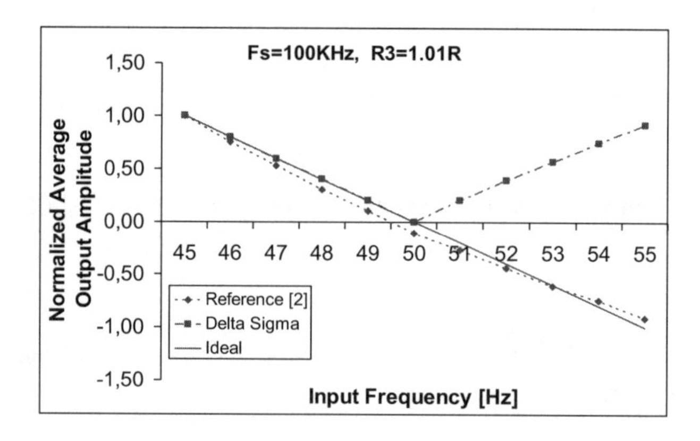


**Fig. 10.16.** Sensitivity of the system when R3 deviates for 1% from an ideal value [10]

If we assume ±1% component tolerances, then both systems are very sensitive to changes in the values of $R_3$ and $C$; Fig. 10.16 shows the result of a simulation for an ideal case and the case when $R_3 = 1.01R$. We can see that both systems are equally sensitive to component tolerances of the analog phase shift circuit.

## 10.6 CONCLUSION

In conclusion, we can say that there are many real-life problems where Δ–ΣM can be successfully employed. In this chapter, we elaborated the problem of direct dynamic measurement and its error estimate. We presented a two-arm bridge for frequency deviation measurement. A two-arm bridge method was based on the use of delta modulation and arithmetic operations on Δ–ΣM pulse streams of two identical synchronous Δ–Σ modulators. The results indicated a good linearity between frequency deviation and pulse count over the range of operation, when Δ–ΣMs are synchronous and properly over-sampled. Component mismatch of Δ–ΣM does not have significant influence on the linearity of the measurement. The phase shift circuit of the system in Fig. 10.7 depends on ω, R and C. Thus, any one of these quantities can be measured, provided the other two are known. It is important to point out that a relatively simple method, proposed in [6], has a serious disadvantage in the presence of noise. Fig. 10.2 may appear to be less costly, but if we add a filter to remove noise from the input signal, the cost of the higher order filter will offset the cost of the comparator based analog to digital converters. The logic circuitry of both methods is identical, thus the cost is the same. Having in mind the present state of VLSI technology and power consumption, implementation of this system should not be a problem for sampling frequencies of the order of MHz.

## REFERENCES


1. M. Freedman, D.G. Zrilic, *Nonlinear arithmetic operation on the delta sigma pulse stream*, Signal Processing, 1990, Vol. 21, pp. 25-35.
2. I. Galton, *Granular quantization noise in a class of delta-sigma modulators*, IEEE Transactions on Information Theory, 1994, Vol. 40, No. 3, pp. 848-859.
3. S. Rabinovich, *Measurement errors: theory and practice*, American Institute of Physics, N.Y., 1993.
4. T. Kasparis, N. Voulgaris, C. Halkias, "A Method of the Precise Measurement of the Difference Between Two Low Frequencies", IEEE Trans. on Instrumentation and Measurement, vol. IM-34, No. 1, March 1985, pp. 95-96.
5. M. Ahmed, "Power system Frequency Deviation Measurement Using an Electronic Bridge", IEEE Trans. on Instrumentation and Measurement, Vol. IM-37, Nol 1, March 1988, pp. 147-148.
6. F. Kay, W. South, "Design of a Power Stabilizer Sensing Frequency Deviation", IEEE Trans. on Power Apparatus and Systems, Vol. PAS 90, No. 2, March/April 1971, pp. 707-711.
7. J. Candy, G. Temes, Editors, "Oversampling Delta-Sigma Data Converters", IEEE Press, 1992.
8. R. Steele, "Delta Modulation Systems", Pentech Press, 1075.
9. N. Kouvaras, "Operations on Delta Modulated Signals and their Applications in the Realization of Digital Filters", Institute of Electronic and Radio Engineers, 1978, pp. 437-438.
10. Dj. Zrilic, "Circuits & Systems for Functional Processing of the Modulated Pulse Density Stream", US patent #6,285,306BI.
11. Dj. Zrilic, "Computation Based on Delta-Modulation Representation of Measurement Results", International Workshop on Applications of Interval Computing, El Paso, TX, 1995, pp.227-231.
12. Dj. Zrilic, N. Pjevalica, "Frequency Deviation Measurement Based on Two-Arm Δ–ΣM Modulated Bridge," IEEE Transactions on Instrumentation and Measurement, April 2004, Vol. 53, No. 2, pp. 293-300.

# CHAPTER 11 LPΔ-ΣM AND BPΔ-ΣM CIRCUITS

## 11.1 INTRODUCTION

Many communication systems use conventional bipolar square-law expansion and compression of voice signals to improve performance. A traditional compander (compressor + expander = compander) employs a pair of analog multipliers with a supporting number of discrete passive components such as dc biased diode, very large capacitors and resistors. These large components prevent a complete monolithic implementation of IC companders. Disadvantages of the traditional analog approach are:

- Bulky and expensive discrete components.
- Sensibility to the parametric mismatches is inherent to discrete components.
- Aging problems over a broad range of environmental conditions for the full life of the product.
- The higher supply voltages required for analog circuits.
- Small drifts in values of passive discrete components can be a source of specification violations.

In summary, non-integrated circuit solutions have difficulties satisfying critical conditions.

This chapter presents both mixed-mode and digital mode compander circuits. Both methods use Δ-ΣM as a basic A/D converter because of inherent ability of the Δ-ΣM to perform division and multiplication. For example, the output binary sequence of Δ-ΣM reflects the ratio between a slow changing input signal and a modulator's reference voltage. Therefore, changing the level of reference voltage, the pulse density at the output of Δ-ΣM is changing as well. This means that the input signal can be scaled by the externally supplied reference voltage. In chap. 6, we showed that the output pulse density stream can be used for multiplication purposes [1]. If a Δ-ΣM pulse stream is used to switch a reference voltage to some multiplying circuit, the resulting output signal is proportional to the reference voltage signal being switched. Examples in chap. 6 illustrate the multiplication of an arbitrary signal with switching the Δ-ΣM sequence, and we have shown in chap. 5 the possibility of direct nonlinear operations on the

Δ-ΣM pulse stream. In this chapter, we will illustrate implementation of both mixed and digital mode compander and demonstrate the possibility of linear arithmetic operations on BPΔ-ΣM pulse stream.

## 11.2 TRADITIONAL APPROACH OF COMPANDING

Non-uniform quantization can be considered to be uniform quantization preceded by compression of the dynamic range of the signal, which has the effect of favoring low amplitudes to the detriment of high amplitudes. The idea is the same as that for analog transmission, the source of noise here being the quantization noise. The original dynamic range (of sampled transmission) must be clearly re-established with respect to the demodulation by means of a strictly reciprocal expansion characteristic as shown in fig. 11.1.

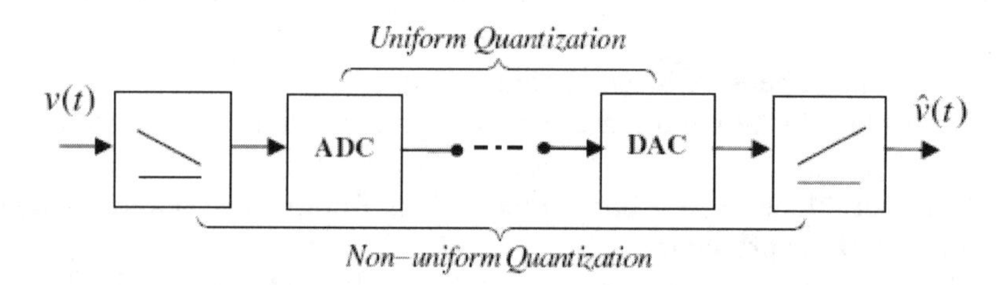


**Fig. 11.1.** Quantization with companding

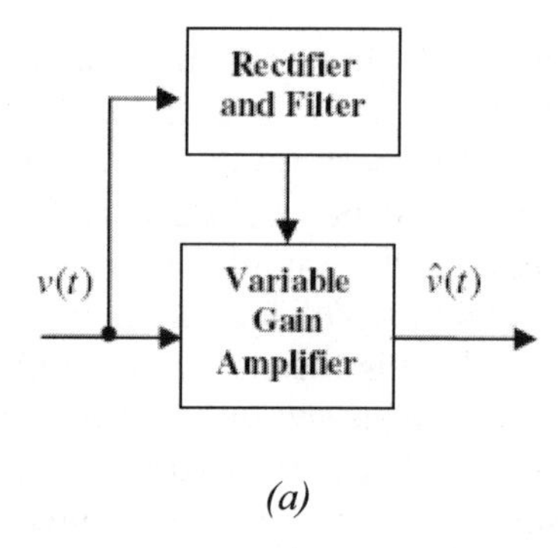


*(a)*

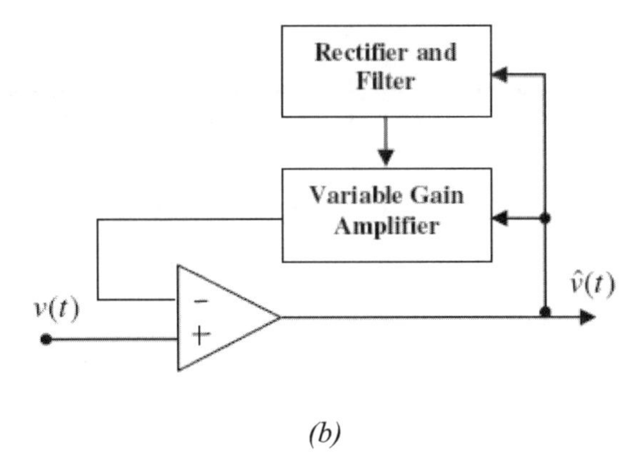


*(b)*

**Fig. 11.2.** (a) Square-law expander, (b) square-law compressor

Traditional expander and compressor concepts are shown in figures 11.2(a) and 11.2(b). Disadvantages of the traditional analog approach of compander circuit implementation are well known. To mitigate existing problems, Takasuka [1] proposed a mixed mode approach based on Δ-ΣM.

## 11.3 A MIXED MODE COMPANDER APPROACH

The implementation of a mixed mode compander circuit using switch capacitors and delta-sigma modulation is described in [1] and [2]. Fig. 11.3 shows a compander circuit configuration based on Δ-ΣM.

The input signal to be compressed is applied to Δ-ΣM. Digital pulse stream $V_D$ is low-pass filtered to get compressed analog signal $V_C$. $V_C$ is then rectified and again low-pass filtered to get nearly DC value for $V_{ref}$. $V_{ref}$ is then switched in a multiplying circuit and fed back into the modulator. For a sinusoid input, the input/output relationship of the compressor circuit can be described as

$$\frac{V_{in}\sin wt}{V_{ref}} \div \frac{V_{in}\sin wt}{V_C}, \frac{V_{in}\sin wt}{V_C} \div V_C \sin wt, or\ V_C = \sqrt{V_{in}}$$

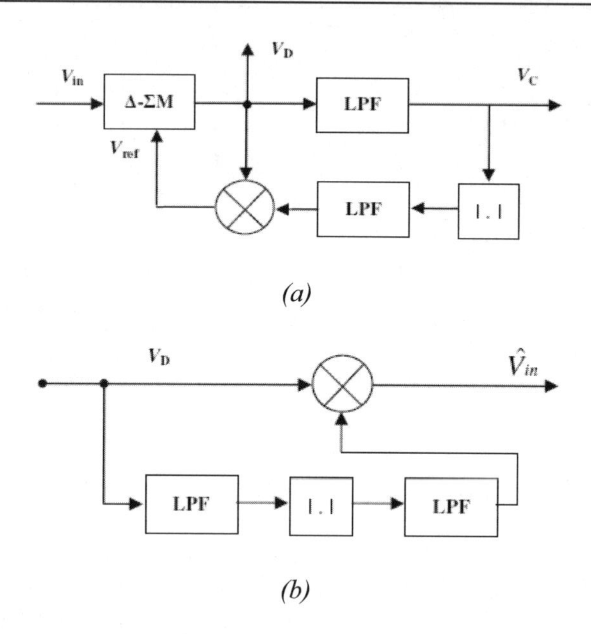


**Fig. 11.3.** Block diagram of Δ-ΣM compander circuit, (a) compressor, (b) expander [1]

where $V_{in}/V_{ref} = V_c$ represents amplitude of the compressed signal. This equation describes a 2:1 compression ratio. It is important to point out that the output of Δ-ΣM reflects the ratio between a slow changing input signal (as compared to the sampling frequency) and the modulator's reference voltage. Changing the level of reference voltage, the pulse density of the output of Δ-ΣM is changing as well. This means that the input signal can be scaled (divided) by the supplied reference voltage. The output pulse density stream reflects the ratio between input and reference voltage. Fig. 11.3b shows the expander circuit. Compressed digital signal $V_D$ is applied to both the low pass filter and switching multiplier. The output of the envelop detector (rectifier + low-pass filter) is a DC signal proportional to the amplitude of the input signal. A pulse amplitude modulation (PAM) output of the switching multiplier is low-pass filtered and the resulting decompressed signal $\hat{V}_{in}$ is shown in fig. 11.4. We can conclude that the Δ-Σ modulator output pulse density stream can be used to switch another voltage reference to the input of some switching multiplier (sample and hold circuit, as described in [1] and [2]).

This sample and hold circuit is an integral part of a one-bit A/D converter implemented with switch capacitors. The resulting output of a DAC is proportional to the voltage reference being switched. As can be seen,

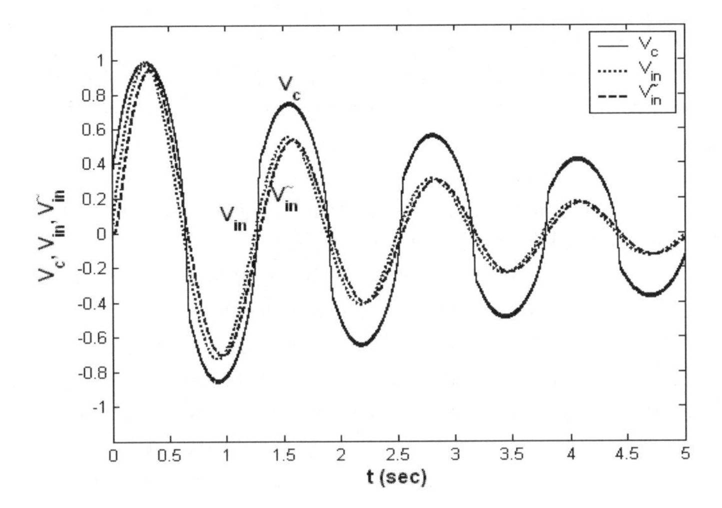


**Fig. 11.4.** Waveforms of mixed mode companding circuit shown in fig. 3

division and multiplication is realized without quadrant multiplication and division as in [1] and [2]. The question is: can we perform direct compression on the Δ-ΣM pulse density stream without converting it into an analog signal back and forth as in the mixed mode approach?

## 11.4 A DIGITAL SQUARE-LAW COMPANDER

Delta-sigma modulation is a popular method for high-resolution analog-to-digital conversion. An input signal is sampled at a frequency much higher than the Nyquist frequency and typical resolution of Δ-ΣM circuits in the market is 20-24 bits [3]. One of the drawbacks of Δ-ΣM is the high over-sampling rate. This results in high storage capacity requirements for uncompressed Δ-ΣM data. This means that compression is inevitable. Since the theoretical limits of lossless compression of one-bit delta-sigma signals are prohibitive, we will be focused on narrow band audio signals where some distortion is permitted.

### 11.4.1 Square-Law Compander

The block diagram of a proposed digital square-law compressor is shown in fig. 11.5(a). The input signal to be compressed is applied to a Δ-ΣM and comparator circuits $C_1$ and $C_2$. Outputs of the comparator are then fed into an AND-OR circuit to get a rectified input signal $A$ and $B$ in digital form, as shown in fig. 11.6. Digital output $C$ of the rectifier configuration is then passed through a digital low-pass filter implemented as in [4], [5], and [6], which represents a Δ-Σ demodulator. The digital output is integrated to get nearly dc signal $E$, which serves as the reference voltage $V_{ref}$ to the Δ-ΣM. The output pulse density stream $V_D$ from the Δ-ΣM represents the ratio between the input signal $V_{in}$ and reference voltage $V_{ref}$.

Fig. 11.5(b) shows a block diagram of an expander circuit. Compressed digital signal $V_D$ is applied to both the low-pass Δ-Σ filter (Δ-Σ demodulator) and the switching multiplier. The output of the Δ-Σ demodulator is rectified as in fig. 11.5(a) and again low-pass filtered to get a nearly dc signal $F$. Output of the switching multiplier, which is a PAM signal, is further fed into a low-pass filter; the resulting expanded (decompressed) signal is $\hat{V}_{in}$ . Fig. 11.7 shows the waveforms of the input signal $V_{in}$, the compressed signal $V_C$, and the expanded signal $\hat{V}_{in}$ . We can see that for higher input levels the compressed signal is clipped (distorted), while for lower input levels (amplitude less than ±1V) the compressed signal is amplified. It is important to mention that Δ-ΣM works properly in our case if the input signal amplitude is limited to or less than 1V.

We have shown that most of the square-law compander circuit can be implemented using the digital, bit-serial technique. The only analog component in our implementation is the first order, low-pass RC filter. The proposed implementation is based on direct Δ-Σ arithmetic operations on a serial pulse density stream obtained from a first order Δ-ΣM.

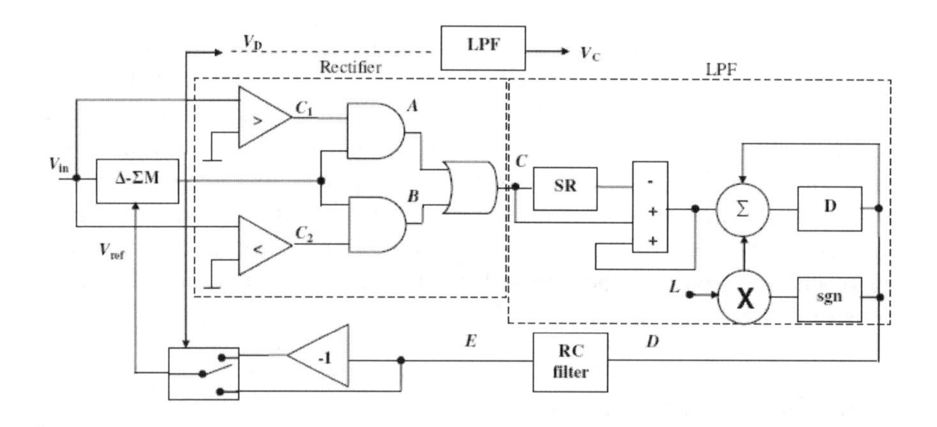


*(a)*

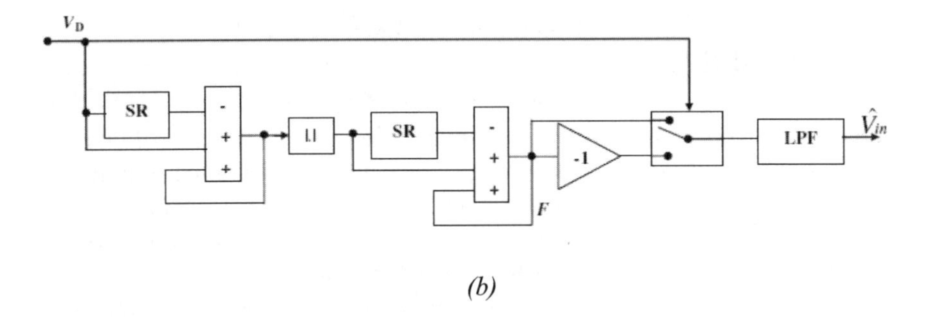


*(b)*

**Fig. 11.5.** Block diagram of a digital Δ-ΣM compander circuit, (a) compressor, (b) expander

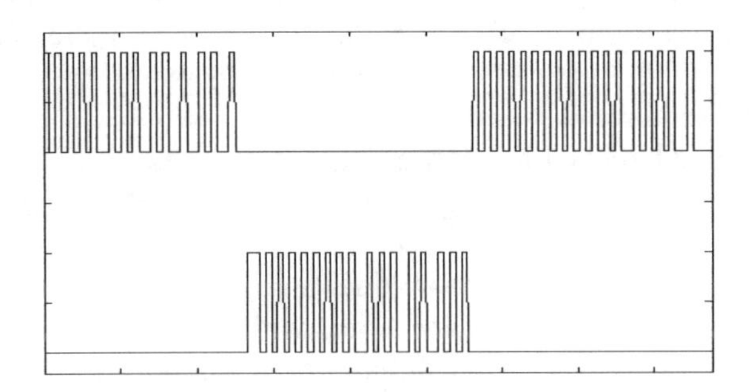

**Fig. 11.6.** Output of the rectifier configuration in the compressor circuit

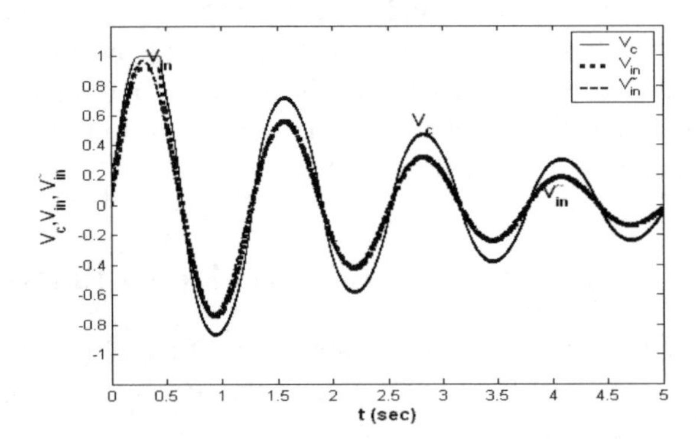


**Fig. 11.7.** Waveforms of the proposed digital compander circuit shown in fig. 11.5

## 11.5 ADAPTIVE LINEARIZATION OF A POWER AMPLIFIER

The next generation of wireless communication systems, both military and commercial, will require considerably lower power dissipation with higher rates. While data rates are increasing to accommodate video, data and voice, a conflicting need exists to reduce power consumption and extend battery life. Modern wireless communication requires reduction in energy consumption of at least one order of magnitude. An important factor is the reduction in battery voltage. The lower battery voltage is crucial to lower power dissipation in the digital components. However, lower voltage for RF circuits means serious challenges in order to maintain power efficiency and linearity of the power amplifier (PA), which consumes maximum power in a RF system. A variety of factors must be addressed to achieve the objective of dramatically lower power dissipation in communication systems. The requirements are:

- For improved battery life and weight.
- Modulation methods that employ the lowest possible amount of energy per bit.
- Devices that are inherently more linear.
- Antennas, power combiners and filters with low loss and small size.
- The trade-off between digital signal processing power efficiency and RF circuitry.
- The need for increased bandwidth, etc. [7, 8].

The greatest leverage for reducing the power consumption of a wireless transceiver is provided by the transmitter's output power amplifier. The power amplifier is currently the "long pole in the tent" as far as cost, power consumption, reliability, and system performance are concerned [9]. In the previous section, we presented the idea of mixed-mode processing of a Δ-Σ modulated pulse stream. In this section, first some existing approaches of adaptive linearization will be presented, then a novel idea of compression using BPΔ-ΣM will be elaborated.

### 11.5.1 Existing Approaches to Adaptive Linearization

High frequency power amplifiers operate most efficiently at saturation, i.e. in the nonlinear range of their input/output characteristics. This phenomenon has traditionally dictated the use of constant envelope modulation methods such as FM or GMSK. However, continuing pressure on the limited spectrum available is forcing the development of spectrally more efficient linear modulation methods such as MQAM and QPSK with pulse shaping. Since their envelopes fluctuate, these methods generate inter-modulation products in a non linear power amplifier. In a mobile environment, restrictions on out-of-band emissions are stringent. The designer is faced with two alternatives, back off an inefficient class A amplifier to an even more inefficient, but with a linear operating region, or linearize the amplifier. Fig. 11.8 shows a generic model for many adaptive amplifier linearization methods [9, 10].

All signal designations refer either to complex baseband signals or to the complex envelope of bandpass signals. The linearizer creates a pre-distorted version $V_d(t)$ for input to the power amplifier. The feedback path directs a portion of the real bandpass PA output to a quadrature demodulator for recovery of the complex envelope. Its output $V_f(t)$ is a scaled, rotated and possibly delayed version of $V_a(t)$. The same oscillator is used in up and down conversion for coherence, and note that a phase shifter is required for stability.

Pre-distortion is the most commonly used technique for linearizing an amplifier [11, 12]. This technique consists of a non-linear process inserted between the input signal and amplifier. This non-linear process generates inter-modulation products that are anti-phase of phase-conjugate to those produced by the amplifier, thereby canceling out the undesired inter-modulation products. This process has the disadvantage of being open looped in nature and is therefore very sensitive to variations in the pre-distortion parameters. Mapping pre-distortion techniques has several drawbacks [13, 14]:

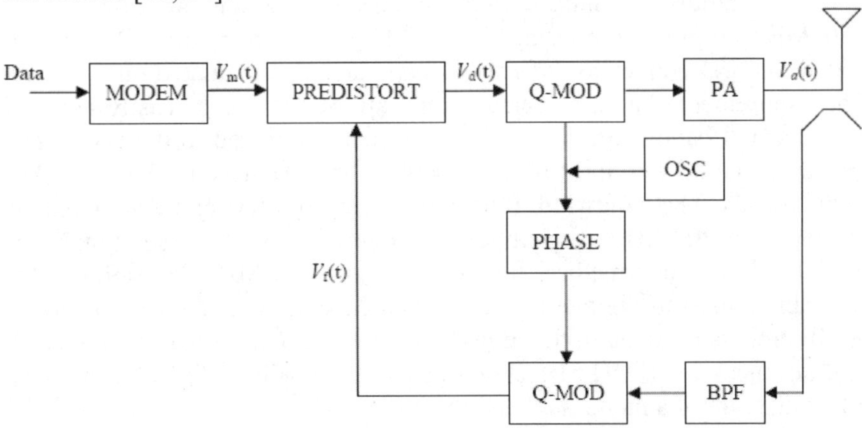


**Fig. 11.8.** Generic configuration for adaptive linearization [10]

1. The look-up table is a 2M word long for 10-bit representation of real and imaginary parts of $V_{in}$, and increases to an 8M word for 11-bit representations.
2. It requires a phase shifter in the feedback path for stability in the adaptive update.
3. Convergence is slow (10s at 16k symbols/s).

Yet another method of adaptive linearization using pre-distortion is described in [15]. This paper describes an adaptive pre-distortion system to linearized class B and C radio frequency high power amplifiers. The system, which can handle a 30 dB power control range, uses a digital signal processing approach with microprocessor and look-up table to pre-distort the baseband signals. The proposed system is capable of handling symbol rates of 8.5k symbols/s using π/4 QPSK modulation. An excellent overview of the methods and concepts of linearization of Power Amplifiers (PA) can be found in [16]. Our approach describes a method of adaptive pre-distortion to linearize power amplifiers using bandpass delta-sigma modulation (BPΔ-ΣM).

### 11.5.2 Basic Idea of Compression using BPΔ-ΣM

Delta-Sigma modulation is well recognized and covered in several books and many articles [17, 18]. Use of bandpass delta-sigma modulation (BPΔ-ΣM) in the switching mode of high efficiency power amplifiers is reported in [7]. The reported simulated amplifier efficiency was in order of 70% for 850 MHz amplification using GaAs HBTs. A bandpass Δ-ΣM is well suited for A/D conversion of narrow band signals modulated on a carrier. The requirement of a high oversampling ratio (OSR) can be easily satisfied as the signal bandwidth is usually very small compared to the center frequency $f_c$. Good examples of such narrow band signals are FM and AM radio signals. The bandwidth (channel spacing) of AM and FM signals is 10 kHz and 200 kHz, respectively. In radio receivers, these signals are modulated in an intermediate frequency (IF) of 10.7 MHz. In brief, a BPΔ-ΣM can be used to digitize the IF modulated signals. The goal is to move the IF processing stage to the digital domain. We believe that with the current advances of RF VLSI technology this is possible. Fig. 11.9 shows a block diagram of a modern receiver using Δ-ΣM [15].

In addition, our goal is to employ the Δ-ΣM signal processing approach not just at the receiver, but at the transmitter as well. Our preliminary research work presents a novel compander architecture that is integrable with other standard transmit/receive signal processing circuits. The method of compression can be seen as a method of pre-distortion in the PA linearization process. Our goal is to elaborate this idea even further. The previously discussed preliminary results have suggested a new solution for the old problem of linearizing power amplifiers (PA). The proposed block diagram is shown in fig. 11.10. Intermediate frequency signal $V_{IF}$ is digitized using BPΔ-ΣM, band-pass filtered and amplitude modulated by a RF carrier wave. Part of the output signal of PA is then demodulated, rectified, and low-pass filtered to get nearly dc signal, which is controlled by digital pulse stream *C*.

Fig. 11.11 shows waveforms of uncompressed and compressed RF signals. Output spectra of PA for both uncompressed (CH2) and compressed (CH1) RF signals are shown in fig. 11.12. Spreading of spectra occurs because uncompressed signal leads PA into saturation.

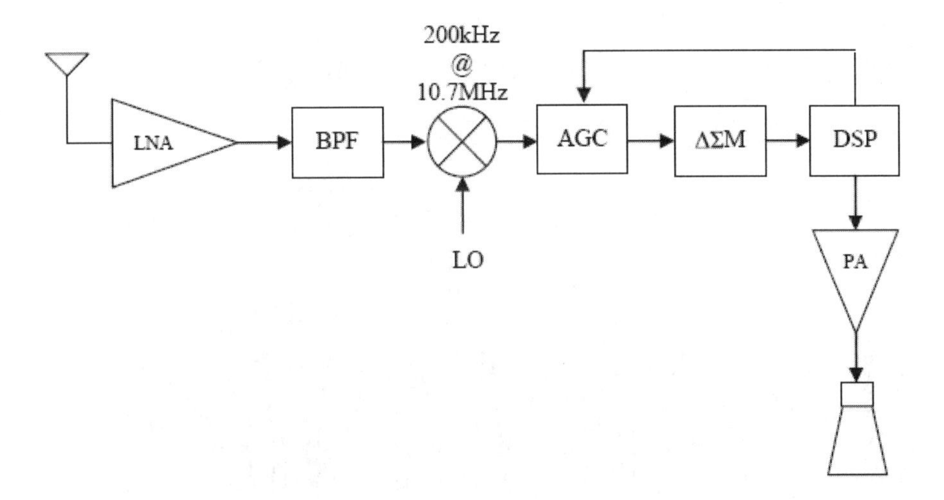


**Fig. 11.9.** Block diagram of a modern receiver using a Δ-ΣM [15]

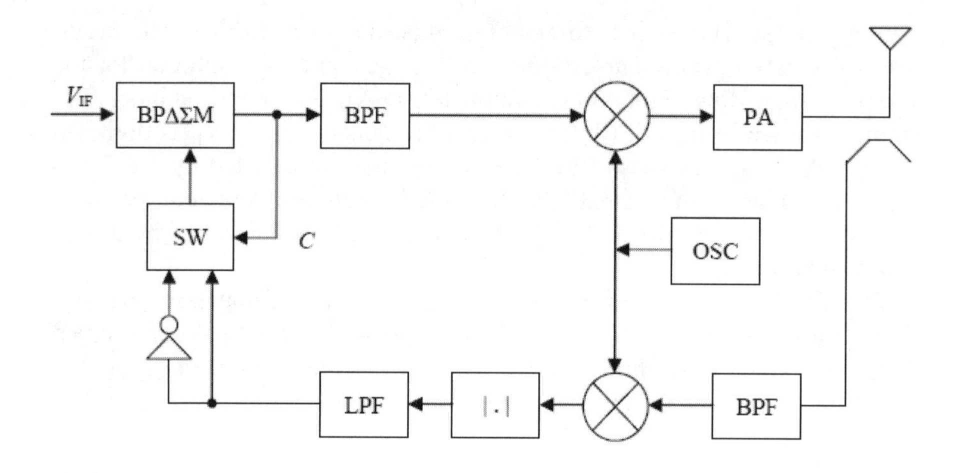


**Fig. 11.10.** Block diagram of proposed compression system [19]

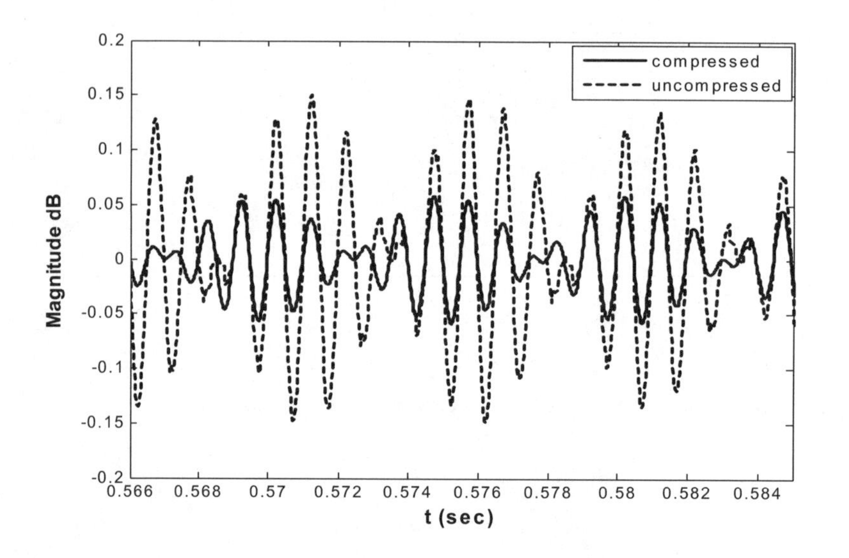


**Fig. 11.11.** RF waveforms of compressed and uncompressed signal

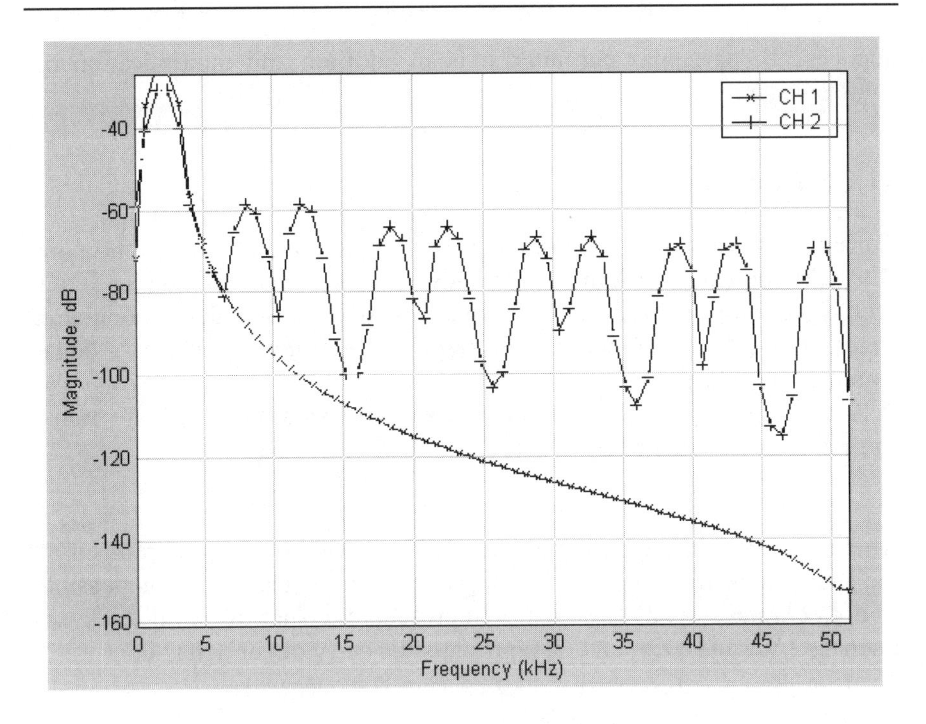


**Fig. 11.12.** Output spectra of saturated (CH2) and unsaturated (CH1) PA

## 11.6 ARITHMETIC OPERATIONS ON BPΔ-ΣM PULSE STREAMS

### 11.6.1 Introduction

A method of direct processing of delta modulated sequences has been presented in chaps. 2 and 4. The required circuit for addition of two binary ΔM sequences is simple and consists of conventional binary full adder and a D flip-flop. In fact, the delta adder is a conventional binary full adder with interchanged sum and carry outputs. The approach of Kouvaras, introduced in chap. 2, leads to the possibility of implementation of digital filters with straight delta modulated input and output signals, which are not intermediately transformed.

The ultimate goal of modern radio communications is to perform A/D conversion at RF level. Usually, BPΔ-ΣM is used to perform this task. The question now is, can we perform arithmetic operations on BPΔ-ΣM sequences? In particular our interest is in addition and multiplication by some constant.

### 11.6.2 Addition

Consider a system, as shown in fig. 11.13, for addition of two synchronous BPΔ-ΣM sequences. We intend to show that the same delta adder (DA), used for addition of two binary sequences of the linear delta modulator, chaps. 2 and 4, can be successfully used for addition of band-pass delta-sigma sequences.

According to fig. 11.13, band-pass signals $x_{bp}(t)$ and $y_{bp}(t)$ are transformed to binary sequences

$$\{X_n\} = \ldots.., X_{-2}, X_{-1}, X_0, X_1, X_2, \ldots..$$

$$\{Y_n\} = \ldots.., Y_{-2}, Y_{-1}, Y_0, Y_1, Y_2, \ldots..$$

where $X_i$ and $Y_i$ take values of +1 or -1. These values are synchronized by the same clock frequency. According to Kouvaras (chap. 2), it is possible to define a new signal $\{S_n\}$, which represents the sum of two discrete sequences $\{X_n\}$ and $\{Y_n\}$, and $C_n$ represents the carry out of delta adder.

$$S_n = 0.5[X_n + Y_n - (1 - X_n Y_n)C_{n-1}]$$

$$C_n = X_n Y_n C_{n-1}$$

$$C_{n-1} = \pm 1,\ n = \ldots, -1, 0, +1, \ldots$$

The terms above take the value of +1 or -1 and thus represent delta modulated sequences. Let the input signal $x_{bp}(t) = \sin\omega_m t \sin\omega_c t$ represent the IF band-pass signal at the radio receiver, where $\omega_m$ and $\omega_c$ are modulating and carrier frequencies, respectively. Let $y_{bp}(t) = 0$ (then $Y_n = I_n$ represents the idle sequence defined as $I_n = \ldots, -1,+1,-1,+1,\ldots$).

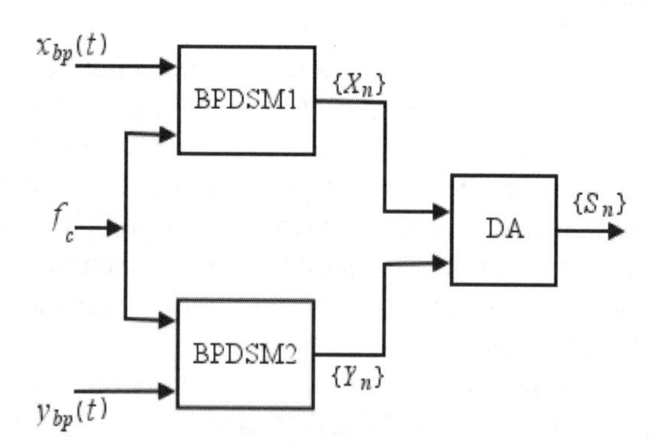


**Fig. 11.13.** System for addition of two BPΔ-ΣM synchronous sequences [19]

Fig. 11.14 represents addition of these two sequences. It can be seen that the demodulated sequence $\{S_n\}$ represents half the amplitude of the sum of input signals $x_{bp}(t)$ and $y_{bp}(t)$. Fig. 11.15 shows the case of addition when $x_{bp}(t) = y_{bp}(t)$. We can also see that by adding the signal with itself, attenuation by the delta adder can be overcome. With the success of addition, we show next how we can multiply a band-pass delta-modulated sequence with a constant $\alpha$, $\alpha<1$, such that $p(t) = \alpha x(t)$.

### 11.6.3 Multiplication

Assume that $\alpha = (0.1101)_2$, then the band-pass delta multiplier has the form shown in fig. 11.16. Fig. 11.17 then presents relevant waveforms for the arithmetic operation of multiplication by a constant $\alpha$=0.8125. We can conclude that simple digital circuits can be used for implementation of digital filters at IF frequencies when BPΔ-ΣM is used as the A/D converter. We hope that this approach will open new possibilities for direct processing of the BPΔ-ΣM pulse density stream. Existing problems of the quadrature sampling method of complex down conversion, such as an exact 90° phase difference between two given oscillators, two ideal mixers, two identical A/D converters, etc. can be overcome using direct mixing and filtering of BPΔ-ΣM pulse stream.

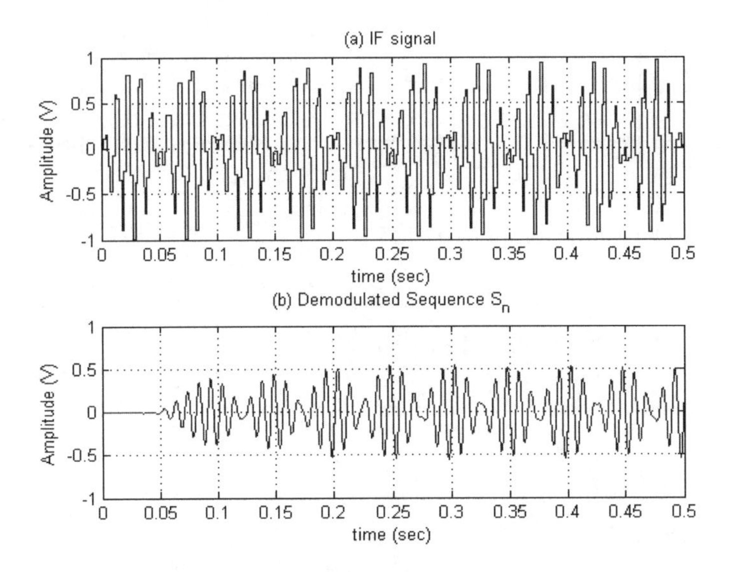


**Fig. 11.14.** (a) IF signal, (b) demodulated sequence $\{S_n\}$

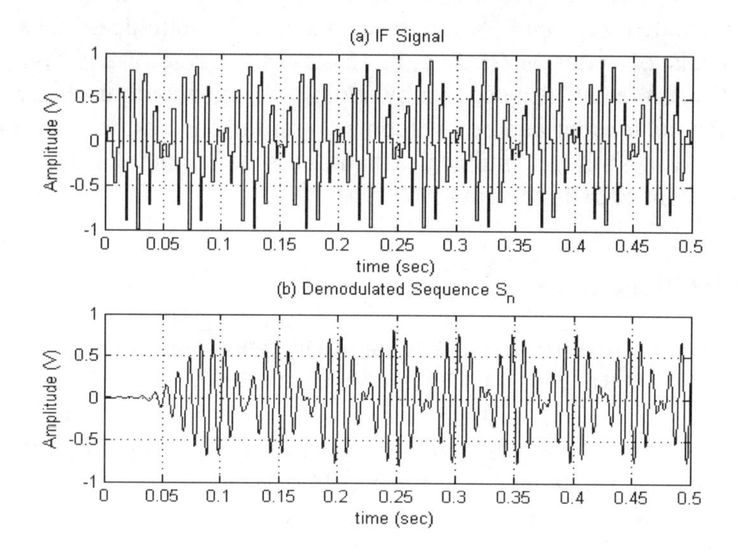


**Fig. 11.15.** (a) IF signal, (b) demodulated sequence $\{S_n\}$ when $x_{bp}(t) = y_{bp}(t)$

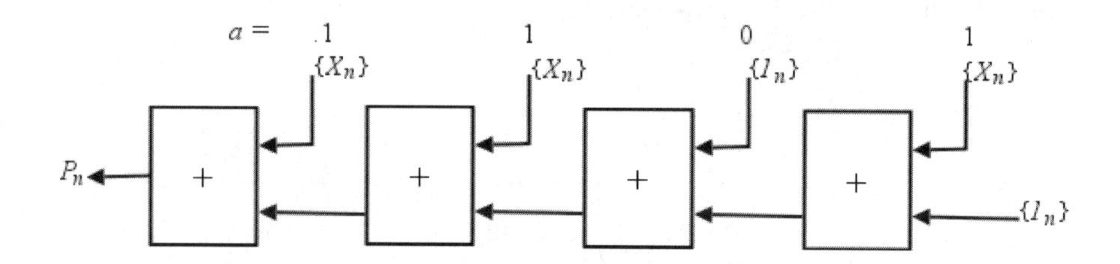


**Fig. 11.16.** Delta Multiplier for $\alpha$=0.8125

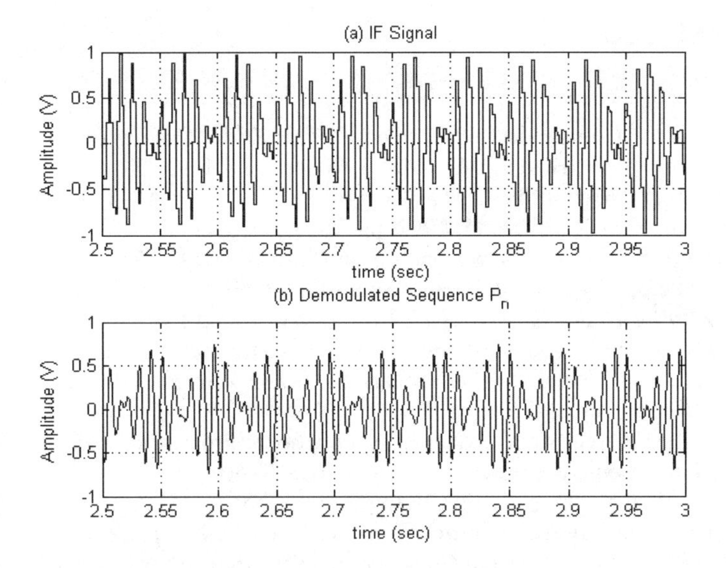


**Fig. 11.17.** (a) IF signal, (b) multiplied signal p(t) = $\alpha x(t)$

## 11.7 CONCLUSION

In this chapter, we have shown a novel approach to implement a square-law compander circuit. The proposed implementation was based on direct arithmetic operations on a serial Δ-ΣM pulse density stream. We have shown that BPΔ-ΣM can be successfully employed in systems for linearization of a nonlinear power amplifier. In addition, we have shown that arithmetic operations on BPΔ-ΣM sequences are possible.

## REFERENCES

1. K. Takasuka, H. Hisajima, K. Takahashi, A. Barlow, "A Sigma-Delta Based Square-Law Compander", *Proceedings of the IEEE Custom Integrated Circuits Conference*, Boston, MA, May 1990, pp. 12.7.1-12.7.4.
2. H. Quiting, "Monolithic CMOS Companders Based on Δ-Σ Oversampling", *Proceedings of IEEE ISCAS*, 1992, pp. 2649-2651.
3. Analog Devices Corp.: www.analog.com/sigma-deltaADCs.
4. Dj. Zrilic, G. Petrovic, B. Yuan, "Simplified Realization of Delta-Sigma Decoder", *Electronics Letters*, 1997, Vol. 33, No. 18, pp 1515-1516.
5. M. Freedman, Dj. Zrilic, "Nonlinear Arithmetic Operations on the Delta-Sigma Pulse-Stream", Signal Processing, *Elsevier Science Publishers*, B.V., Vol. 21, 1990, pp 25-35.
6. Dj. Zrilic, "Circuits and Systems for Functional Processing of Delta Modulated Pulse Density Stream," US Patent #6285306BI.
7. P. Asbeck et al, "Device and Circuits Approaches for Next Generation Wireless Communications," *Microwave Journal*, Feb. 1999, pp. 22-34.
8. J. Kenny, A. Leke, "Design Consideration for Multicarrier CDMA Base Station Power Amplifiers," *Microwave Journal*, Feb. 1999, pp. 76-84.
9. S. Cripps, "RF Power Amplifiers for Wireless Communications," *Artech House*, 1999, ISBN 0-89006-989-1.
10. J. Cavers, "Amplifier Linearization Using a Digital Predistorter with Fast Adaption and Low Memory Requirements," *IEEE Transactions on Vehicular Technology*, Vol. 39, No. 4, Nov. 1990, pp. 374-382.
11. J. Namiki, "An Automatically Controlled Predistorter for Multilevel Quadrature Amplitude Modulation," *IEEE Transactions on Communications*, Vol. COM-31, May 1983, pp. 707-712.
12. T. Nojima, T. Konno, "Cuber Predistortion Linearizer for Relay Equipment in the 800 MHz Band Land Mobile Telephone System," *IEEE Transactions on Vehicular Technology*, Vol. VT-34, Nov. 1985, pp. 169-177.
13. A. A. Saleh, J. Salz, "Adaptive Linearization of Power Amplifiers in Digital Radio Amplifiers," *Bell System Technical Journal*, Vol. 62, No. 4, April 1983, pp. 1019-1033.
14. Y. Naguta, "Linear Amplification Technique for Digital Mobile Communications," *Proceedings of the IEEE Vehicular Technology Conference*, 1989, pp. 159-164.
15. M. Faulkner, M. Johannson, "Adaptive Linearization Using Predistortion – Experimental Results," *IEEE Transactions on Vehicular Technology*, Vol. 43, No. 2, May 1994, pp. 323-332.
16. S. Kenney, "Methods and Concepts for Power Amplifier Linearization," *IEEE Radio and Wireless Conference, RAWCON 2001 Workshop*, Walthom, Mass., USA, Aug. 19-22, 2001
17. J. Candy, G. Temes, Editors, "Oversampling Methods for A/D and D/A Conversion," *IEEE Press*, New York, 1991, ISBN 0-87942-285-8.

18. J. Van Engelen, R. Van Plassche, "Bandpass Sigma Delta Modulators," Kluwer Academic Publishers, Boston 1999, ISBN 0-7923-8698-1.
19. Dj. Zrilic, "Circuits and Systems for Linear and Nonlinear Processing of Band-Pass Delta-Sigma Pulse Density Stream", Patent Disclosure, November 18, 2004.

# Index

Printing: Krips bv, Meppel
Binding: Litges & Dopf, Heppenheim